Editor: Lt. Col. David Eshel (IDF ret.)
Editorial Contributor: Bryan Perrett

Special assignments: Brenda Ralph Lewis (SOE)
Tamir Eshel (Special Forces, Rangers, SEALs)

Editorial coordinator: Lawrence D. Rifkin
Design: John Forster

Published by Arco Publishing, Inc.
215 Park Avenue South, New York, NY 10003

Library of Congress cataloging in Publication Data

Eshel, David
Elite Fighting Units

1. Commando troops. 2. Parachute troops. 3. Marines. 4. Special forces (Military science). 5. Guerrillas.
I. Title.
U262. E84 □ 1984 355.3'5 84-7419

ISBN 0-668-06206-1

Color: Reshet Kav, Migvan

Printed in the Netherlands by Van Boekhoven-Bosch b.v., Utrecht

ELITE
FIGHTING UNITS

ELITE
FIGHTING UNITS

David Eshel

ARCO PUBLISHING, INC.
New York

PREFACE

If one had to choose one action from among many with which to illustrate the fighting spirit of elite forces it would be difficult to better the remarkable fight put up by Lieutenant Kieth Mills and his twenty-two Royal Marines at Grytviken, South Georgia, on 2 April 1982 – a fight which came within an ace of completely wrecking the Argentine invasion plan.

Mills had been told to make the Argentines fight for British territory and to resist for perhaps half an hour. In private, he made his attitude to his orders quite clear: 'The hell with half an hour – I'm going to make their eyes water!' He consolidated his position near the harbor and rejected a surrender call from the enemy invasion force, which consisted of a large transport escorted by one of the Argentine Navy's new A-69 corvettes. Three helicopters, a Puma and two Alouettes, promptly flew in to land troops. The Puma, with twenty men aboard, was riddled with 600 rounds of rifle and machine gun fire and crash-landed on the opposite side of the harbor; none of its occupants were seen to emerge. Next, an Alouette was seriously damaged and took no further part in the action. Meanwhile, the enemy corvette had steamed cockily into the harbor to give the landing party close gunfire support. The Marines opened up with everything they had and a later Argentine count revealed no less than 1275 strikes with small arms ammunition. Worse, an 84mm anti-tank round skimmed off the water and blasted a hole through the ship's side; two 66mm anti-tank rounds struck her forward turret, jamming its elevation; and a second 84mm round hit the corvette's Exocet launchers, but by the vilest ill-luck this round did not explode. Had it done so, fire would have raged throughout the ship and the whole operation would probably have been cancelled. The warship hastily turned tail, but continued to give support from a respectful distance outside the harbor. By now the Argentines had about 50 men ashore and their fire, coupled with that of the corvette, kept the tiny party of Royal Marines pinned down. A fierce fire-fight raged on but at length Mills, his retreat blocked by sheer cliffs, negotiated a surrender. He and his men had caused an incredible amount of damage and inflicted casualties beyond their own numbers, at the cost of one Marine wounded. They had fought for over an hour and the eyes they had made water belonged to the commander of the landing force, Captain Alfredo Astiz – the notorious 'Captain Death' – whose earlier mood of relaxed optimism had changed to one of depression.

At first the Argentines approached Mills' group with something like awe, refusing to believe that they were the sole garrison; they sincerely believed that they had been fighting a much larger force, supported by artillery and anti-aircraft weapons. Some were generous in their praise and one officer told Mills, 'You took on two ships, 500 Marines and three helicopters! There are no kamikazes left in Japan – they're all here!'

Lieutenant Mills, aged 22, was subsequently awarded the Distinguished Service Cross – an award second only in merit to the Victoria Cross – for the highest standards of junior leadership displayed in this action. He and his party were repatriated via neutral Montevideo, but returned to the South Atlantic to seek, and find, their revenge.

Over the years, the term elite has undergone a subtle change in its meaning. Historically, it is usually associated with Imperial or Royal Guard formations, the ranks of which were filled with soldiers of proven ability, and which would normally be committed during the climax of a set-piece battle to administer the final crushing blow. The age of total war has inevitably witnessed the disappearance of such *Gardes du Corps* as decisive battlefield instruments, although the contribution made by those that survive is invariably well above average.

Today, the term is generally applied to two types of units. The first fight as infantry and are normally committed to only the most hazardous of operations. Because of the nature of these, or their method of arrival on the battlefield, they require men of above-average physical fitness, determination and initiative who are trained to a high standard of battlecraft. Such units include, *inter alia*, the Royal Marine Commandos, the United States Marine Corps, the French Foreign Legion, the parachute formations of the world's major armies, and mountain troops. The second type of elite unit tends to specialise in clandestine operations of various types, the underlying principle being that a small group of highly trained men frequently achieve better results than could be obtained by a much larger force, or results which simply could not be obtained in any other way. Once described as unorthodox, this approach is nothing of the kind. Unorthodoxy arises when a unit is employed in a manner which is contrary to its purpose and training; these units are trained specifically for clandestine warfare and covert operations, which have long formed part of the pattern of 20th Century conflict. There are, in fact, a number of historical precedents for units of this type, which are sometimes known collectively as Special Forces. One of the most notable was Rogers' Rangers, raised by the British in 1755 for service in the forest wilderness during the French and Indian Wars in North America. Another was the famous Corps of Guides, raised on the North-West Frontier of India by the then Lieutenant Harry Lumsden, its purpose being not only to provide trustworthy guides for troops in the field, but also to *'collect intelligence beyond as well as within the border.'* Lumsden cared nothing for his men's race, religion or background, but was extremely selective in his choice of recruits. The men he wanted were 'notorious for desperate deeds, leaders in forays, who kept the passes into the hills and lived amid inaccessible rocks.' In an age when armies still fought in scarlet and blue, the Guides dressed in khaki for the sake of concealment, the first unit to do so. As to their fighting ability, they had few, if any, equals on the Frontier. When, in 1879, the British Residency in Kabul was attacked by an armed mob numbering several thousand including mutinous Afghan troops who brought up their cannon, the small garrison

of 75 Guides under Lieutenant Walter Hamilton, VC, died to a man. But before the last shot was fired they had killed 600 of the enemy and wounded up to twice that number. Their memorial at the Regimental Depot at Mardan is inscribed as follows: 'The annals of no army and no regiment can show a brighter record of devoted bravery than has been achieved by this small band of Guides.'

It is, of course, possible to quote similar historical parallels, but there can be no doubt whatever that Rogers' Rangers and the Guides must be numbered among the most honorable ancestors of today's Special Forces, including the British Special Air Service Regiment and the United States Army's Green Berets. The qualities demanded are the same: above average physical fitness, stamina, intelligence, imagination, initiative, self discipline, compatibility and iron determination.

The First World War and its immediate aftermath provide few examples of the sort of operation now mounted as a matter of course by the world's elite forces, the reason being that flexibility and originality at the tactical level were firmly suppressed by the rigid military concepts of the time. There were, however, two notable exceptions. The first was a British amphibious operation which took place on 22 April 1918, the intention being to block the exits from the U-boat bases of Zeebrugge and Ostend with concrete-filled blockships. This was only a qualified success, but the following year Royal Naval CMBs (Coastal Motor Boats) penetrated the Russian naval base at Kronstad, sinking two battleships, a cruiser and a submarine depot ship, and disabling two destroyers. This effectively destroyed Soviet naval power in the Baltic, which Lenin had publicly declared to be a 'Russian lake'. These dramatic results were obtained by eight CMBs, each armed with one or two 18-inch torpedoes and machine guns and manned by two young officers and a mechanic, and the raid thus epitomizes the Special Forces' philosophy of using small, highly trained specialist groups to inflict massive damage on an enemy who feels secure behind his defenses.

The stage was now set for the techniques to be further developed during World War II.

THE SECOND WORLD WAR
THE UNITED KINGDOM

In the days following Dunkirk it was realised by the British War Cabinet that the only way in which the enemy could be hurt was by means of a series of raids against a German-held coastline which stretched from the North Cape to the Pyrenees. In an earlier century such a policy had been described as 'breaking windows with guineas'; the remark was political in origin and the speaker failed to see that the military consequences extended well beyond the damage caused and the casualties inflicted. Damaging raids, if made often enough, compelled the enemy to look to his own defense and move large numbers of troops – who could be more gainfully employed elsewhere – into the threatened area; they also kept him on edge, upset his civil population, and made trouble for his political establishment.

In 1940 the need for Great Britain to hit back quickly was of paramount importance. It seemed entirely logical that such a raiding force should be formed by the Royal Marines, but to assemble sufficient suitable men would have meant stripping the Fleet. That would have taken time which was simply not available. Instead, volunteers were drawn from every branch of the Army and quickly formed into battalion-sized units. Some measure of the urgency with which the project was viewed can be gauged by the fact that the first

raid against the coast of France took place during the night of 23/24 June, just three weeks after the completion of the Dunkirk evacuation and only 19 days after the unit involved had assembled.

The purpose of these units was mirrored in their title of Commando, which was taken from the small, elusive and hard-hitting Boer groups which had tied down 250,000 British troops during the South African War of 1899-1902. Their original internal organisation of ten troops each of three officers and 47 other ranks bore comparison with that of Rogers' Rangers and other locally raised light troops which the British Army had employed in North America in the Eighteenth Century.

The Commandos worked under the con-

Hardelot – April 1942; a raiding party from 4 Commandos, along with a Canadian contingent, raids the French coastal village of Hardelot from 6 LCA. Commanding the operation was Lord Lovat, later distinguished leader of the 1st Special Service Brigade, among the first ashore at Normandy on D-Day, 1944.

Return from Abercrombie; the commandos are transported back on an LCA (Landing Craft, Assault).

trol of the Chief of Combined Operations, whose responsibility it was at this stage to plan and execute operations which would hit the enemy hard and in the most spectacular way possible. The first major raids, carried out in 1941, established the Commandos' reputation. On the Lofoten Islands, at Spitsbergen and at Vaagso, they set vital oil storage tanks ablaze, wrecked gun positions, destroyed communications equipment, inflicted sharp casualties and returned home with prisoners for interrogation. On 28 March 1942 they attacked the port installations of St Nazaire, their specific target being the gates of the Forme Ecluse, the only dry dock on the Atlantic coast of France capable of taking the German battleship *Tirpitz*. Escorted by a flotilla of motor launches, the destroyer HMS *Campbelltown*, her bows

packed with explosives, rammed the gates under intense fire. Commandos swarmed ashore to indulge in an orgy of destruction. Fierce fighting raged throughout the dock area and the town but by dawn the Germans believed that the raid had been repulsed and began rounding up prisoners. At noon the *Campbelltown* blew up, shattering the lock gates and killing a party of senior German officers who had boarded her; further delayed charges continued to rock St Nazaire throughout the day. The cost of the raid had been high, but it had achieved its purpose.

The Dieppe raid, mounted by the 2nd Canadian Division on 18/19 August 1942, was also costly, but provided essential information concerning the enemy's coastal defenses which was put to good use when the Allies landed in Normandy in June 1944 with unexpectedly low casualties. At Dieppe, the losses would have been even higher had not the heavy coastal artillery batteries north and south of the town been eliminated respectively by Numbers 3 and 4 Commandos in a brilliantly planned and ruthlessly executed operation.

There were, too, many smaller but equally significant raids on the coast of Occupied Europe and against Axis territory in the Mediterranean, including the gallant failure of the attack on Rommel's North African headquarters. The effect of these raids boosted British morale and forced the Wehrmacht to deploy more divisions in Norway and France than strict necessity dictated – divisions, moreover, that were sorely needed on the Russian front. There is ample evidence that the problem caused Hitler acute anxiety. The long Fuehrer War Directive No. 40, issued on 23 March 1942, is as much concerned with the defeat of local raids as with the prospect of a major Allied landing in the west. By October 1942 the activities of the Commandos had driven him into one of his maniacal rages and on the 18th of that month he issued his famous Extermination Order, otherwise known as Directive 46A. This required that British 'sabotage troops,' whether in uniform or not, whether armed or not, were to be 'killed to the last man in battle or in flight,' or, if captured indirectly, were to be handed over to the SS with, presumably, similar terminal consequences. Some men were indeed shot after being captured but the majority of German commanders chose to ignore the directive either because they were conscientiously opposed to it or because they saw that such a radical departure from the accepted customs of war would inevitably provoke retaliation.

By now the Commandos had become an international by-word for efficiency, dash and dedicated courage. Their green beret symbolised their incredibly tough selection and training procedures, the Commando knife their remorseless determination to win. The balance of victory was now swinging in favor of the Allies and the role of the Commandos was beginning to change from that of raiding force to spearhead of amphibious assault landings; the arm as a whole, too, had been strengthened by the addition of several Royal Marine Commandos. During the last years of the war the history of the arm was, to some extent, the history of the war itself, and among the campaigns and battles in which the Commandos fought are the following: Sicily, Salerno, Lake Comacchio, D-Day, Normandy, Walcheren, Holland, the Rhine Crossing, North-West Germany, Burma, the Arakan and Kangaw.

Wounded but happy survivors of the Dieppe raid on the quay side at Newhaven.

19 August 1942: a naval motor launch with four of the landing craft used in the raid on Dieppe.

Royal Marine Commando Cap badge.

S.B.S. COMMANDO

V. COMMANDO

SPECIAL AIR SERVICE

2ND S.A.S.

PARACHUTE REGIMENT

① Paratroop wings.

② Paratroop Regiment Cap badge.

③ Pegasus Shoulder Flash.

SAS jeep patrol; North Africa, 1944.

Normandy, 1944;
Royal Marine Commandos coming
ashore. They carry
with them small
motorcycles.

A further aspect of amphibious warfare was also undertaken by a number of small volunteer units, the functions of which are now carried out by the Royal Marine Special Boat Squadron. During World War II they went by various titles including the Small Scale Raiding Force, the Royal Marine Boom Patrol Detachment, 101 Troop, and the Combined Operations Assault Pilotage Parties (COPPs). The role of these units was the clandestine gathering of information concerning the suitability of beaches for assault landings and close reconnaissance of the enemy's coast defenses, their normal method of arrival being by submarine, which surfaced briefly to allow them to float off in their canoes, and would then pick them up at an agreed rendezvous on completion of the mission. This sort of operation was carried out regularly along the coast of Occupied Europe, in the Mediterranean, on the coasts of Burma and Malaya, and latterly also produced invaluable intelligence prior to the Rhine and Irrawaddy river crossings.

The RMBPD also had another function, which was the destruction of shipping inside enemy harbors. One such raid, which took place in December 1942, involved paddling over 60 miles up the Gironde river to the port of Bordeaux, where limpet mines were attached to several Axis blockade runners which were reported to be on the point of sailing for Japan with the latest German radar and radio equipment aboard. The raid, which formed the subject of the film Cockleshell Heroes, seriously damaged four

British commandos
landing at Normandy;
D-Day.

ships, but most of the participants were captured and shot out of hand. In June 1944 the RMBPD scored another spectacular success when a raiding force penetrated Porto Largo harbor on the Greek island of Leros and damaged two Italian destroyers which the German Navy could have put to good use. The ships were towed to Greece for repair, where they were sunk by the RAF.

The Royal Navy, too, had developed means of penetrating the enemy's most secure defenses and inflicting crippling damage. These consisted of two-man Human Torpedoes and four-man midget submarines known as X-Craft, both of which would position themselves under the target vessel and jettison heavy explosive charges and limpet mines; these would be fired by time fuse while the submarine made its escape.

The first Human Torpedo attack took place in January 1943 against shipping moored in Palermo harbor, sinking the newly completed Italian cruiser *Ulpio Traiano* and an 8500-ton transport, the *Viminale*. In June 1944 the 8-inch gun cruiser *Bolzano*, which had fallen into German hands when Italy had requested an armistice, was sent to the bottom in La Spezia by the same means.

In September 1943 several X-Craft attacked the battleship *Tirpitz* as she lay at anchor in Alten Fjord, Norway, and damaged her so severely that she was out of action for months. In April 1944 X24 entered Bergen and sank the 7500-ton supply ship *Barenfels*, returning in September of that year to drop her charges beneath a newly arrived and urgently needed floating dock, which split in two and sank. Each of these attacks required

4th Special Service Brigade HQ positioned at St. Aubin-sur-Mer, Normandy. The men are supported by a Churchill tank.

1st Special Services Brigade commandos dig in after having relieved elements of the 6th Airborne Division outside Ranville, Normandy; 6 June 1944.

BRUNEVAL 1942
CAPTURE OF THE RADAR

RAF Bomber Command had striven to carry the war to the German homeland since the beginning of World War II, a goal which became even more imperative as Britain, isolated by the occupation of France, was forced on the defensive. Her only remaining offensive option was aerial bombardment, until sufficient strength could be assembled to invade the mainland, a goal that seemed ages away in the winter of 1941/42.

To mount an effective strategic bombing offensive, however, the RAF would have to overcome *Luftwaffe* air defense measures which were rapidly growing in efficiency as new electronic means were fielded. The German radars, although late in coming, were surrounded by mystery, with little reliable information being available to British intelligence. *Luftwaffe* radar sites were springing up all along the occupied European coast, and by October 1941, some 27 stations were identified, spotted mainly by recce Spitfires, stretching from Bordeaux in France to Norway. RAF aircraft had earlier detected radar signals while flying over enemy territory, and efforts were made to determine their exact frequency. However, sufficient details were still lacking as 1941 drew to a close. A photograph taken by Chinese agents in Berlin actually showed a dish-shaped structure mounted on a flak tower situated in the Berlin Zoo – closer examination later identified it as the new Wuerzburg radar that was enter-

THE RAID ON BRUNEVAL

Bruneval; the target – an isolated villa housing the radar site HQ later assaulted by Major Frost and his team. The photo below was brought back by Flt. Lt. Tony Hill on his second Spitfire recce sortie, and played a vital role in the planning of the raid.

The official briefing map for the operation.

ing service. But the hunt for details pertaining to the radars deployed in the *Luftwaffe* forward defensive zone was now becoming top priority; time was running short if the 1942 spring bomber offensive was to be implemented with acceptable loss rates.

Late in November 1941, Dr. Charles Frank, examining a medium-level air photograph of the *Luftwaffe* Freya station near St. Bruneval, noticed a foot trodden track leading from the cliff edge, where the bowl-shaped antenna was sited, to an isolated building. The photo was one of a series taken on routine recce sorties flown by the Photographic Reconnaissance Unit at RAF Benson, Oxfordshire. Due to the photos having been taken from too great a height, the details were very difficult to define. But the sharp-eyed scientist managed to make out a particularly interesting object near the house, which he thought to be part of an antenna. Calling his superiors, he recommanded a special low-altitude photo sortie to get a more detailed coverage.

It was decided to call in some of the

most experienced recce pilots from Benson and prepare a special low-level mission. Flight-Lieutenant Tony Hill, DFC, was selected and took his Spitfire PR IV over Cap d'Antifer. Sweeping low over the cliffs, he passed exactly over the radar site; alas, on his return, he found to his dismay that his cameras had malfunctioned. He managed, however, to take a quick look at what he described as an electric 'bowl-fire' – shaped object. Next morning, Hill broke the unwritten law and flew the same sortie again, this time taking what was to become one of the most famous low-level photos of the war. Tony Hill's magnificent oblique photograph was so sharp that it furnished excellent detail for intelligence researchers.

The idea of pilfering the Bruneval radar first took shape in the minds of Combined Operations HQ staff, headed by Naval commodore Lord Louis Mountbatten, in January 1942. After examining the available photographic data, it was suggested that a parachute raid should be mounted instead of the then-usual commando attack with troops being landed by sea. Although the Wuerzburg radar site was less than a hundred meters from the coast, the steep cliffs overlooking the beach, on which the

Germans had situated strong defenses, ruled out a seaborne landing. Although the command post seemed to be in the isolated house, clearly visible in Tony Hill's photo, it became known to the planners that, some 150 meters to the north, a farmhouse called Le Presbytère sheltered about a hundred men, part of a coastal defense company manning the outposts, as well as the off-duty signallers and radar operators. More serious was the company-size local reserve force with armored cars, located at St. Bruneval village, about three kilometers to the south. A secondary macadam road led from Bruneval to the north, about one kilometer from La Presbytère, with dirt tracks leading towards the coast. A regional reserve, an infantry regiment, was in garrison near Le Havre to the south.

With such a defensive force in the operations area, it would be a touch and go affair, the success of which would depend on thorough planning and ultra-rapid execution. A highly skilled and trained unit was called for to carry out the mission within a well integrated combined arms operation.

As the objective for the whole action was specific, the British scientists, headed by Dr. R.V. Jones, set about considering the points in which their special

Flt. Lt. Tony Hill, Photo Recce Unit, RAF Benson. He was killed a year later flying a recce sortie over Le Creusot.

Paratroopers dropping from an RAF Whitley bomber converted for airborne missions.

interest would lie, defining the exact parts which would be required in order to gain maximum intelligence. The best solution, of course, would be if the complete radar station was dismantled and brought back to Britain. However, this was ruled out completely, as the German station was much too heavy and large to lift and transport manually to the beach. The 'boffins' therefore designated the priority parts to be brought back. To ensure sufficient technical knowledge, an RAF radar expert was called for and a volunteer came forward. Flight Sergeant E.W.F. Cox was thoroughly briefed for his task, undergoing special parachute training at the Ringway training center, making five jumps to qualify for the coveted blue parachute wings.

Meanwhile 'C' company, 2nd Parachute Regiment started its thorough training, the men still unaware of the job that lay ahead. As a cover, the men were told they were preparing for a demonstration to be mounted before the Royal Family. Training was exact and strenuous, with everyone working long hours under the eye of Company Sergeant Major G. Strachan, formerly from the Black Watch Regiment, an untiring and impressive figure of a man.

As the training proceeded, the officers were at last let in on the secret, viewing their target for the first time at Medmenham, where the photographic section of the RAF had prepared a special model of the Bruneval site – an exact reproduction of Tony Hill's photo made by a peacetime sculptor, Flight Lieutenant Geoffrey Deeley, who worked with a staff of specialists. On this model and with vertical aerial photos of the objective, Frost and his officers worked out their plan in detail.

In all, 119 officers and men were to take part, divided into three assault parties, each codenamed after famous sailors. It was to be a unique combined service operation, actually the first mounted on such a scale.

Designated to fly the assault teams to

Men of C Company, 2nd Parachute Reg., training in England for the jump.

their objective was No. 51 Squadron RAF, flying Armstrong-Whitworth Whitley bombers and commanded by the famous Wing Commander P.C. Pickard, DFC, already a well-known bomber pilot.* To bring the assault teams back to England was the task of the Royal Navy, which designated a special naval force for the job, including assault landing craft and motor gunboats supported by two destroyers. Aboard the ALCs were a mixed party of the Royal Fusiliers and South Wales Borderers, whose duty it was to cover the withdrawal of the paratroopers as they scrambled down the cliffs.

The Bruneval mission gained urgency and momentum on 11 February when the German battlecruisers *Scharnhorst* and *Gneisenau* weighed anchor from their moorings at Brest and, heading out to sea, defied both the RAF and Royal Navy, dashing in broad daylight through the Channel, jamming British radar from as close as 20 miles offshore. The blow to British morale was potentially devastating and urgent action was required – Churchill himself urged that the operation proceed without delay.

On 15 February the force made its final parachute drop over Salisbury Plain, and by the 23rd all was ready.

The raiding force was split into three parties, aimed at landing together on a designated drop zone far enough inland to avoid immediate detection, but allowing rapid approach to the target area with well-defined ground markers visible in the dark.

Commanding the main party was Major J.D. Frost. It consisted of fifty men

Paratroopers making ready before entering their aircraft, a Whitley bomber.

Wing Commander Pickard, CO of 51 Squadron, examining a captured German helmet brought back by the paratroopers.

subdivided into two groups, one to assault the radar, the other the isolated house. Included were radar expert Flight Sergeant Cox and a party of sappers from 1st Para Field Sqn. RE, under Captain

* Group Captain Pickard, one of the most highly decorated RAF bomber pilots of the war, was to lead the famous Mosquito raid against Amiens Prison, where he was killed on 18 February 1944.

Denis Vernon, whose task it was to dismantle the radar sections pointed out by Cox. The sappers were specially trained on a British artillery radar set, which came closest to what the boffins guessed the German equipment looked like.

Leading the second party with forty paratroopers was Lieutenant Charteris, whose task was to secure the defended

Company C paratroopers returning home on a Royal Navy assault craft.

Major John Frost, DSO, MC, commander of the Bruneval raid.

beach area and cover the retreat. The last party, commanded by Lieutenant John Timothy, with 30 men, was to act as a blocking force, sealing off the area from advancing German reinforcements.

After several postponements due to bad weather, the stage was finally set for the mission on Thursday, 27 February, with a full moon and high tide combining to create excellent weather conditions, the target area covered by snow. Twelve twin-engined Whitleys lined up on the tarmac at RAF station Thruxton in Wiltshire as the paratroopers, all in high spirits, started to emplane with their

cumbersome gear. One by one the heavily-laden aircraft took off, Major Frost flying in Pickard's plane.

As the Whitleys roared overhead aiming for the French coast, the naval task force was already well underway, making for the night rendezvous. To distract the enemy's attention, Bomber Command had mounted several diversionary raids north and south of the target area and the force arrived over the dropping zone just after midnight, encountering some flak from the batteries around Le Havre; some aircraft were hit, but little damage was caused. However two Whitleys taking sharp evasive action came slightly off course and dropped Lt. Charteris' party not quite in the drop zone.

Major Frost jumped first, quickly followed by his party, and all arrived safely, landing quietly in the snow precisely at the designated spot, completely unnoticed by the sleepy Germans, who were used to aircraft noises and took no interest in the intruders. Operation 'Biting' was on.

As the sound of engines faded into the night, Frost whispered final instructions to the party leaders around him and set off for the assigned objectives. Moving quickly with four men at his heels, the major burst into the villa, killing a German guard with automatic fire. Outside, sounds of battle were heard as Lieutenant Young and his men raced into the radar position. As Frost joined him the area was already clear; five of the six Germans were dead and a terrified prisoner cowered in shock as he watched Captain Vernon and Cox inspecting the equipment. Soon the sappers were working feverishly

to dismantle the vital parts, directed by flashlamp which started to draw fire from the Germans who were wide awake in Le Presbytère. Time was running short; motor noises were heard from the east as German reinforcements started to arrive. They were fired on by Lt. Timothy's road block party in a bid to win time.

Major Frost urged Cox and Vernon's sappers to hurry, as the fire became more determined. By now the raid had been in progress for fifteen minutes and an urgent retreat was necessary to get to the beach before the Germans arrived in force. Units were ripped off their consoles, manhandled by crowbars; hauling the precious prize on their backs, the party started to withdraw towards the beach, meeting more persistent fire as they went. The coastline was still densely defended by alert Germans, as Lt. Charteris' party had arrived late due to the faulty drop. Racing at top speed, the party arrived on top of the cliffs at the last moment and stormed the German defenses just as Frost and his withdrawing party were about to mount their own assault.

Just after two in the morning, with the beach secure, the parties assembled and waited for the naval evacuation to commence. Vernon's sappers had achieved a near miracle, dismantling almost all the priority parts asked for.

Several attempts to contact the offshore navy failed, but then the boats came in, opening covering fire on the clifftop just as the first Germans started to arrive. Frost quickly directed the embarkation, and soon all were aboard the assault craft, which started backing away out of range of the German fire.

On board a telecommunications expert examined the captured equipment, finding it highly interesting.

The Bruneval raid was totally successful. Most of the vital radar equipment was captured, and three prisoners, among them a qualified German radar operator, were taken against the loss of two dead, seven wounded, and six missing.

The Bruneval Wuerzburg radar find was to have far-reaching effects on Bomber Command operations and electronic warfare in the war. The Germans themselves were most impressed with the operation, and praised its efficiency. Major Frost was to gain fame again, leading the 2nd Parachute Regiment at Arnhem Bridge in 1944, and as a Major General became one of Britain's most distinguished soldiers.

British airborne patrol at Oosterbeek, on the outskirts of Arnhem, Holland.

Men of the Glider Pilot Regiment mop up German snipers in a deserted Arnhem building. Soldier near the entrance holds a Sten SMG.

1st Airborne Division dropping on the outskirts of Arnhem. Note the gliders that have already landed at the site.

the strongest nerves and iron determination, but that made by XE3 in July 1945 against the 10,000-ton Japanese cruiser *Takao* was particularly courageous. The *Takao* was lying in the Johore Strait, between Singapore Island and the mainland of Malaya, and XE3 made her approach through shallow water and on a falling tide to drop her charges beneath the enemy ship and mine her hull. Throughout the most critical phase of the operation the X-Craft was in serious danger of being crushed between the cruiser's keel and the sea bed. The *Takao's* bottom was blown out and she settled into the mud, never to sail again. XE3's commander, Lieutenant Ian Fraser, and her diver, Leading Seaman Mick Magennis, who had placed the limpet mines and then left the boat for a second time to clear an obstruction that was inhibiting her escape, were both awarded the Victoria Cross.

Prior to World War II the United Kingdom did not possess an airborne arm but the success of German parachute and air-landing operations in the early years of the war made it imperative that when the British Army went over to the offensive it should con-

tain an airborne element. Altogether, three different types of unit were developed:
1. The Parachute Regiment, recruited initially from volunteers drawn from every branch of the Army, whose function it was to secure a strategic objective by dropping onto

British paratroop officer in a forward position near Arnhem; 1944.

On the ground a few moments later.

British paratrooper patrol erecting a roadblock near Arnhem; 1944 Lower left – a PIAT anti-tank launcher; near rail crossing – a soldier with BREN gun in position.

Major General Orde Wingate, founder and leader of the Chindits.

Chindits advance in the Burmese jungle. Facing page: Chindits 'handling' a railway bridge in Burma.

to date, and here too its dash and tenacity won its members the nickname of The Red Devils, conferred by an enemy who was himself capable of offering the most determined resistance imaginable.

In the spring of 1943 the 1st Airborne Division was formed, initiating a growth in British airborne forces that was to be maintained until the end of the war. Further airborne operations, involving both parachute and air-landing troops, took place in Sicily in 1943, in Normandy and Holland in 1944, and during the Rhine Crossing in 1945. The majority were outstandingly successful, but it is the very gallant fight against tremendous odds at Arnhem which is best remembered and which has come to symbolize the very essence of Para toughness, dedication and courage.

In Burma the Chindits, who took their

or close to it, then holding the position until relieved by ground troops.

2. Air-landing regiments, which retained their own identity and which arrived by glider or transport aircraft to support the Parachute troops within the secured objective.

3. The Chindits, formed for deep penetration operations in Burma, who either marched to or were air-landed at their operational base and were then supported by air supply drop throughout their mission.

The Parachute Regiment's first major success was the raid on the Bruneval radar station on 22 February 1942. In a model operation the 2nd Battalion's C Company, commanded by the then Major John Frost, dropped close to the station, which was quickly seized together with parts of the top secret Wuerzburg radar set, and was then evacuated safely by the Royal Navy. The acquisition of this equipment enabled the Allies to develop the necessary electronic counter measures (ECM) with which to defeat the enemy's radar defenses.

The regiment's first operational drops in brigade strength took place in Algeria and Tunisia during November 1942. Thereafter, it fought as conventional infantry for the remainder of the Tunisian campaign. Here it earned nine of the 28 battle honors it has won

name from the stone *Chinthe* or guard dogs found outside Burmese temples, did all that was asked of them, inflicting casualties, severing Japanese communications and absorbing the attention of enemy troops who were desperately needed at the front. The major Chindit penetration, intended to support British operations at Imphal and Kohima, was, however, diverted to assist Lt-General Joseph Stilwell and his Chinese divisions in their attempts to capture Myitkyina. As a result of the indirect pressure applied by the Chindit columns the town eventually fell, but the weeks of gruelling toil in the jungle, tropical diseases and combat had all taken their toll, so that when the Chindits eventually emerged the health of many men was so broken that they were unfit for further military service. The Chindit concept was the brainchild of Major General Orde Wingate, who

was killed in an air-crash during the campaign, and was to find a renewed application during post-war counter-insurgency operations.

The Special Air Service Regiment was born during the war in North Africa and its founder was Major David Stirling, a former Scots Guards officer who had transferred to the Army Commandos. While recovering from injuries received during a parachute drop, he conceived the idea of a deep penetration force of saboteurs which would operate against enemy airfields, inflicting damage out of all proportion to the number of men involved. His plan was approved and he was allowed to raise and train a unit for this pur-

Back to Europe. SAS sniper hunter patrol at Geilenkirchen; November 1944. Note the modifications made for cold weather operations, such as the foldable windshield, and more-standard weapons than those used in the desert.

SAS returning from 3 months of patrol behind enemy lines in North Africa; 1943.

Members of SAS operating with Resistance fighters behind German lines; Italy, 1945.

LRDG (Long Range Desert Group) on patrol in Libya.

SAS officer in 'field' uniform (a traditional Arab head dress and standard British desert uniform).

pose, taking the Winged Dagger – which some suggest was originally intended to represent the legendary sword Excalibur – as its emblem, together with the motto *Who Dares Wins*.

By the time the Desert War had ended the SAS had destroyed almost 400 Axis aircraft, and much else besides. For much of this period they worked in conjunction with the Long Range Desert Group, which had been formed by another far-sighted officer, Major Ralph Bagnold, in June 1940. Bagnold was a noted desert traveller and he recruited like-minded individualists whose common store of knowledge was put to good use. Operating from oases well south of the main battle area, the LRDG used the vast empty spaces with which they were familiar to penetrate far into the enemy's rear, there to carry out covert surveillance, active reconnaissance, beat up the odd convoy, snatch prisoners, drop the SAS raiding parties within marching distance of their objectives and pick them up after they had done their work. The basic equipment of the LRDG was the 1½ ton Chevrolet truck, armed with several automatic weapons, while in later operations, both in North Africa and in Europe, the SAS used the Willys jeep, bristling with machine guns. In the desert both groups dressed to suit the environment, the SAS particularly so, their normal uniform consisting of an Arab burnous, bush shirt, shorts and sandals; during opera-

SAS troop before a long-range patrol. Note the amount of supplies carried and the immaculate jeep – on its return, it sure looked different; Libya, 1942. The jeeps have a special cooling device for hot desert conditions.

tions shaving was abandoned to conserve water, with the result that most men sported piratical beards.

Flexibility is the keynote of the SAS approach and as the war in Africa drew to its conclusion the regiment embarked on a campaign against the Axis garrisons of the Greek islands, using schooners and camouflaged caiques with which to mount their raids. Such was the success of this that the Germans were forced to commit substantial numbers of troops, including a crack mountain divi-

virtually destroyed the logistic infrastructure of the 5th Panzer and 7th German Armies, operating against railways and ambushing road transport far behind the enemy's lines. Similar operations were carried out in the Low Countries and in Germany itself for the remainder of the war, provoking yet another furious outburst from Hitler: 'These men are dangerous. They must be hunted down and destroyed at all costs!' This time his *diktat* was heeded and there is ample evidence that SAS prisoners were savagely tortured before being murdered in cold blood.

Schooner used by the SAS to support operations against Germans in the Aegean islands.

Early morning chill and a road watch behind German lines in Libya – LRDG personnel.

Stirling greets a patrol on its return after three months behind enemy lines.

SAS patrol; Tunisia. Note the many supplies needed for desert survival. They will last for only a week, maybe less. Resupply had to be done by air or other infiltration techniques.

sion, in an attempt to contain the deteriorating situation.

The validity of the concept was now beyond doubt and the strength of the SAS rose to two regiments. It operated in Sicily, Italy and was active in France some time before the D-Day landings. During the Normandy fighting, with the aid of the French Resistance, it

SPECIAL OPERATIONS EXECUTIVE

On 5 June 1944, thousands of Resistance workers were on stand-by in France, Holland, and Belgium awaiting radio signals from London announcing D-Day and the imminent Allied invasion of Europe. They were armed for the coming struggle with rifles, Sten guns, pistols, bazookas, explosives, and other sabotage materials and weapons. Their countries were networked with secret radio transmitters – 150 sets in France alone. Hundreds of rail lines, roads, viaducts, bridges, telecommunications centers, and other targets were earmarked for attack as soon as London gave the word. In fact, the 'secret armies' of the Resistance comprised a cauldron of potential pandemonium primed to pour death and disruption upon the Germans from within Occupied Europe while the Allied invasion assaulted it from without.

This cauldron was set simmering by Special Operations Executive, an intelligence/espionage organization which, at its inception in 1940, faced such enormous problems that it seemed the least likely to score success at so complex a task. For a start SOE roused fierce jealousies among existing intelligence groups whose work it overlapped. It was even regarded as subversive and a suspect foreign influence by expatriates of German-occupied countries who foresaw its agents upstaging native resistance forces. It was loathed with particular intensity by the Free French and their leader, Charles de Gaulle, who saw SOE as just one more sign of the perfidy of Albion. British Service Chiefs fiercely opposed the organization on several counts, one being its mainly civilian make-up. Although SOE employed numbers of men in uniform, such as Major-General Sir Colin Gubbins, Director of Operations, and Colonel Maurice Buckmaster, Chief, French Section, its personnel were mainly civilians who, in peacetime, had been merchants, accountants, actors, writers, professors. Service departments were horrified that this curious crew should be permitted to wage war and rival the armed forces for scarce materials and transport.

Especially odious was the fact that SOE business was very dirty business, employing every filthy trick known to the world of cloak-and-dagger: spreading false rumors, bribery, prostitution, blackmail, forgery, robbery, smuggling, and, of course, sabotage. This wholesale dabbling in the disreputable

Train derailed and set ablaze during the Resistance action in the Saône Valley with the loss of 1.2 million liters of aircraft fuel.

laid SOE wide open to charges of chicanery and crime; its critics then (and even now) did not hesitate to daub its image with every act of double-dealing it was possible to commit. However, the dismal fact was that on 19 June 1940, when SOE was first proposed by Winston Churchill, its underhand methods were the only ones by which the war against Germany could be actively pursued. After Hitler's swift shock conquests in Scandinavia and Western Europe, the armed forces of Britain, the only surviving challenger, had to assume a defensive role. Only SOE remained as a force with freedom of initiative.

The creation of SOE was sanctioned by the British War Cabinet on 22 July 1940. Its many purposes included contacting and training resistance workers in enemy-occupied territories; organizing reception committees for agents; drops of arms, supplies, and money; planning and carrying through acts of sabotage; building up secret armies to aid an eventual Allied invasion; and collecting economic, military, and political information.

Easier said than done, however. The fall of Western Europe had all but destroyed pre-war British intelligence networks and had cloaked the continent in silence and unkown hazard. In 1940, no one knew where potential resistance existed, where agents could find safe sites for landing by plane or parachute, where to conceal equipment, or where to go to ground when in danger of discovery. The situation was later remedied, but only after pioneers had ventured in 'blind'. For survival, the first SOE agents sent to occupied Europe – three parachuted into France in May 1941 – had little more than their wits and their luck to rely on.

Wit, if not luck, had to be of a special order for a man or woman to qualify for work with SOE. The prime requirement was that agents should be fluent in foreign languages and at all times behave like a native. They had to possess talent for organization, the ability to withstand tension and pressure, and the diplomacy required to mediate between warring factions within Resistance movements.

Those who fit the bill were sent for training at special SOE schools set up in Britain, North Africa, Italy, and the Far East. By 1944, 60 such training schools were turning out graduate spies and saboteurs – about 7,500 for Western Europe and some 4,000 for other areas of SOE operations. Here, student agents received instruction in unarmed combat, shooting, dismantling and assembling weapons, parachuting from aircraft, standing up to brutal interrogations, and smuggling on their persons or in their baggage espionage aids such as hollowed-out matchsticks containing microfilms or codes and ciphers reduced to speck-size microdots. They learned how to operate a wireless and practised memorizing an elaborate system of codes and call signs. They were also shown how to use escape aids, like leaf-thin 5 inch steel saws for slicing through prison bars, tiny compasses concealed in buttons, and the ultimate aids of all, the 'L' pill filled with potassium cyanide for use when the last resort was the only resort. Finally, agents had to assimilate the false identity and background concocted for them by SOE.

Before departing on missions, agents underwent meticulous body and clothing searches to eliminate giveaway signs like pencil stubs, cinema or train tickets, labels, and anything else that might betray British origins. Once cleared, agents went to 'holding camps' where they received the last checks and the last briefings from their personal Conducting Officers, and waited for a

Special Operations Executive:
A member of the Maquis checks munitions parachuted by the SOE into France near Chartres.

night dark enough and the aircraft that was to land or parachute them on their missions.

The first mission to France by Lysander, the gawky single-engined 'spy plane' used for landing and pick-up operations, took place on 4 September 1941 – an agent was landed in a field near Issodun and another was collected. The Lysander, painted matt black, spent two minutes on the ground.

By this time, SOE agents had been operating in France for four months. The pioneer trio of Georges Begue, a radio operator, and

Resistance member in France listens for instructions on his radio, supplied by the SOE. His companion keeps watch.

Pierre de Vomecourt and Roger Cottin had parachuted into the Dordogne during May 1941 and had formed the first SOE circuit in occupied France, codenamed 'Autogiro'. There followed drops of more agents and canisters containing weapons in September and October. The agents set about establishing their *reseaux* (nets) both in German- and Vichy-occupied France, setting up their radio posts in 'safe' houses and relaying back to London news of their progress.

Naturally, extreme danger was inherent in this furtive, clandestine work. Agents lived in permanent danger of betrayal by traitors, informers, or other agents broken by torture. They might always be on the edge of making the one simple mistake – for instance mumbling an English phrase while dozing – that would give them away. In fact, out of 480 agents sent into France, 130 were captured, and of these all but 26 perished in gun battles, under torture, in experimental surgical oper-

ations, by firing squad, or in concentration camps.

Dangers especially insidious were German infiltration and 'reversal' of the *reseaux,* and the overcentralization of circuits which enabled the Germans to lop off the leadership at one scoop. After the *reseaux* were regionalised in 1943 and began operating more independently, there was less peril that the detection of one would lead, house-of-cards-like, to the same fate for others.

At this juncture, Resistance work was also improved by greater availability of aircraft, including American Liberators and Flying Fortresses. The result was a dramatic rise in drops of agents and supplies. Sorties over France rose from 101 during September-December 1943 to 609 during January-March 1944, and 1,728 during April-June of that year. In August 1943, No. 138 Squadron dropped 66 agents, 1,452 containers, and 94 packages on 184 missions. Six months later came a 12-week period when the French Resistance received 76,290 Sten guns, 16,945 rifles, 27,961 pistols, 3,441 Bren guns, 572 bazookas, 160 mortars, and several tons of explosives and ammunition.

This influx of arms fully equipped 20,000 and part-armed another 50,000 men, with a convulsive effect on sabotage operations. In one month alone – 25 October through 25 November 1943 – 3,000 acts of sabotage against the French rail system took place, and other attacks left a trail of mangled locomotives, power stations, and transformers, collapsed pylons and cables, and silenced German radio stations. One notable success was the sabotaging of the German radio transmitter at Quatre Pavillons, near Bordeaux – when that was blown up, the Germans lost their main line of communication with U-boats marauding in the Atlantic.

Several times, the activities of French *reseaux* like Pimento, Stationer, Wheelwright, and Acolyte, among others in occupied Europe, formed the stuff of which movie thrillers are made. On 3 June 1943, Brian Rafferty, an Irish SOE agent, masterminded the volcanic explosion at the Michelin works at Clermont-Ferrand, which consumed hundreds of tons of tyres and destroyed the plant in a holocaust of flame. The same summer, SOE agents and Resistance workers sabotaged the Peugeot motor works at Sochaux, the third most important industrial target in France. The series of mammoth detonations which echoed round the factories, where turrets were produced for German tanks, achieved within a few minutes what months of Allied bombing raids had failed to accomplish.

The deadliest weapons production of all was crippled when SOE agents and the Norwegian Resistance, Milorg, sabotaged the Germans' efforts to develop an atomic bomb. In February 1943, the installations and apparatus used in the production of deuterium oxide (heavy water) at the Vemork hydro-electric plant were badly damaged by a network of charges laid under the noses of German guards. It took several months to repair the major damage and restart production. Then, on 20 February 1944, 15,000 liters of heavy water went to the bottom of Lake Tinnsjo after agents planted bombs aboard a ferry. In neighboring Denmark during 1943, SOE and Resistance forces worked in circumstances of utmost danger to relay information back to London about the rocket research station at Peenemunde where the Germans were developing and test-firing the V-I and V-2 'vengeance' weapons. The result was the plastering of Peenemunde by the USAAF and RAF on 18 August 1944.

Operations of this caliber (and there were many more like them) made reality of Churchill's original hope that SOE would 'set Europe ablaze'. However, when Europe finally took fire with Operation Overlord and the D-Day invasion, SOE performed a 'star turn' that eclipsed even the most spectacular of its previous exploits.

Towards the end of 1943, when the run-up to the invasion had already begun, SOE was placed under the control of the Allied Supreme Command, and the following Spring, several dozen agents and radio operators were sent to France. In England, some one hundred 'Jedburghs' – three-man liaison teams – and units of the neo-Commando Special Air Services stood by to follow them once the D-Day landings had taken place. Also on stand-by, twice a month, were radio operators of the Resistance, awaiting radio signals that would tell them of the approach of D-Day. For the event itself, two pre-arranged personal messages were to be relayed to each Resistance group by the British Broadcasting Corporation in London. The first signified 'Stand By – Invasion Imminent', the second signalled the order to start sabotage operations.

On the night of 5 June 1944, five hundred messages like 'The dice are on the table' or 'It is hot in Suez' buzzed over the wires to over fifty underground radio stations, and signalled a tumult of sabotage and destruction in France and the Low Countries. Explosions wrecked rail junctions, lines, tunnels, and bridges. Telegraph, telephone, radio and other communication links were cut. Power stations were blown up, snuffing out lights and silencing factory machinery.

Jedburgh team wireless operator.

After D-Day on 6 June, the Resistance were led by Jedburgh teams who acted as leaders and instructors in ambushing and shooting up columns of troops making for the invasion site in Normandy. Likewise, the Belgians and the Dutch aided the Allies by stunting in similarly obstructive fashion the Germans' attempts to beat back invasion and preserve their Fatherland.

In promoting underground warfare against the Germans to the bonanza proportions of 1944, Special Operations Executive gave civilians a chance never previously offered to them in war on an organized scale – the chance to strike directly at those who oppressed them and do their own bit in person to avenge their wrongs.

GERMANY

Tempting as it might be to regard certain sections of the German armed forces, such as the Waffen ss and the Army's Grossdeutschland Division, as being elite troops, the term can only be applied in the older, traditional sense that would be used in connection with an Imperial Guard. Indeed, these formations were initially raised as an emotional substitute for the vanished Prussian Guard regiments and fought in an entirely traditional manner, using none of the techniques which are the hall-mark of units which are regarded as being an elite in the modern sense. Again, in the final analysis the very existence of 'favored' divisions was ultimately counterproductive, since these creamed off the best men and the lion's share of equipment which could have been put to better use by the Army as a whole.

However, a genuine elite existed in the form of the Airborne troops, both parachute and air-landing, which operated under Luftwaffe control. In April 1940 they were used with imagination, dash and aggression in Norway and Denmark, achieving complete tactical surprise and overwhelming the defenders before they could offer serious resistance. The following month a specially trained group of *Fallschirmjaeger* under Captain Walter Koch landed silently by glider on top of the allegedly impregnable Fort Eban Emael, the lynch-pin of the Belgian defense system and, using prepared charges, grenades and flamethrowers, compelled the garrison to surrender after a stiff fight. Simultaneously, parachute and air-landings were seizing vital bridges and airfields inside the zone declared by the Dutch General Staff to be Fortress Holland, holding on until relieved by fast moving armored columns which had broken through the frontier defenses. The effect of these operations was to paralyse the enemy's command system, pre-

British commandos in Norway.

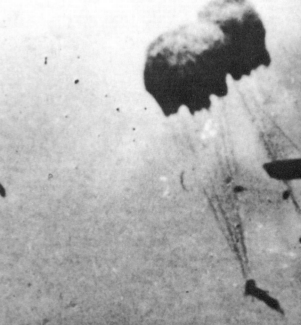

vent a junction between the French and Dutch armies, and knock Holland out of the war in a matter of days.

It was, however, Operation Mercury, the airborne invasion of Crete, that was to bring both triumph and tragedy to the *Fallschirmjaeger*. Many senior German officers doubted whether the operation was viable, but General Kurt Student, the airborne forces commander, was confident that it was. In the event the paratroopers had to fight like fiends simply to preserve a toe-hold on the island, and one complete battalion was wiped out as it landed. Nonetheless, incessant dive-bombing gradually prised loose the British hold, and a misunderstood order led to a needless withdrawal which allowed German reinforcements to be flown in. Thereafter, the remainder of the island was overrun with the assistance of the 5th Mountain Division, a descendent of a much older elite, the *Tiroler Kaiserjaeger*, in which the Austrian royal family had once served.

German casualties had been staggering, amounting to 33% of those involved in the fighting; one in four of the *Fallschirmjaeger* were dead. This, coupled with the heavy loss of aircraft, led Hitler to forbid any more such operations and for the remainder of the war

A German paratrooper jumps a typical low level 'German style' (head down) jump from a JU-52 aircraft.

German paras jump from a JU-52 during the assault on Rotterdam, Holland.

the parachute regiments fought as extremely tough infantry in Tunisia, Italy, France, North-West Europe and on the Russian Front. The last drop ever made by the Hunters from the Sky took place in 1944 during the Battle of the Bulge and was a complete failure, the 800 men involved being so widely dispersed on landing that the Americans were able to round them up without difficulty.

Germany also possessed her own Special Forces unit, the *Bau- und Lehrkompanie* (Construction and Training Company) *Brandenburg*, which was eventually expanded to divisional strength, its members being known as Brandenburgers. The function of this unit was infiltration, causing panic and confusion behind the enemy's lines, and seizing important objectives within the tactical zone by *coup de main*. In this field they performed much useful work, although they never seem to have developed a feel for deep penetration missions in the British manner. Their most notable achievements took place in 1940 when, dressed in civilian clothes or in Dutch or Belgian uniforms, they attempted to capture the more important bridges over the Maas. Some attempts ended in high farce, but others succeeded brilliantly and a grateful Fuehrer conferred no less than 92 Iron Crosses on a single company.

Another unit with a mixed history was ss Parachute Battalion 500, raised in 1943 for special missions, half its members being ss volunteers, the rest being drafted in from penal battalions. Its first mission, an attempt to capture Marshal Tito from his headquarters in Bosnia, was a failure and resulted in it's being cut to ribbons by the Yugoslav partisans. After a spell in the line in Kurland it came under Sturmbannfuehrer Otto 'Scarface' Skorzeny's command, was re-designated ss Parachute Battalion 600 and made up to strength with English-speaking volunteers. Wearing American uniforms and driving captured American vehicles, it was employed in the Brandenburger role during the Battle of the Bulge and for a while caused considerable confusion until the Americans devised their own simple counter-measures. These consisted of questions the answers to which would be known only to Americans, and those of Skorzeny's men who failed the test were promptly despatched in accordance with the accepted usages of war.

Yet despite the uneven record of German Special Forces, they had their dramatic successes, none more so than Skorzeny's rescue of the Italian dictator Mussolini from his mountain prison of Campo Imperatore on 12 September 1943. The troops involved belonged to the Parachute Lehr battalion, reinforced by a small ss contingent. One party travelled overland and captured the funicular railway station at the foot of the mountain, thus severing communications with the resort; the second landed swiftly and silently by glider on the plateau surrounding the hotel. The startled guards offered no resistance and Mussolini was spirited away in a Fieseler Storch light aircraft.

British commandos infiltrating in Norway.

THE SOVIET UNION

During the 1930s the Soviet Union built up the largest airborne arm in the world and in 1938 possessed six fully trained airborne brigades, numbering 18,000 men. Unfortunately, the potential of this force was blunted by Stalin's ruthless purge of the Russian officer corps and completely eliminated for a while by crippling losses in aircraft incurred as a result of Operation Barbarossa, the German invasion of Soviet Russia. At this period, too, the Naval Infantry Corps possessed neither experience nor any real potential, and although both branches of the service would in due course achieve elite status, they first had to learn their trade in the hard school of experience. Indeed, the majority of attempts to employ these forces against the Germans ended in complete disaster, and several of the successes claimed barely find mention in the German records.

At the end of December 1941 the Black Sea Fleet effected an amphibious landing, with paratroop support, at Feodosiya in the Crimea, hoping to relieve pressure on Sebastopol, then besieged by General von Manstein's 11th German Army. After a week's heavy fighting the landing force was virtually wiped out and the survivors forced to take to their boats.

By September 1943, however, the Germans had been placed firmly on the strategic defensive and the Black Sea Fleet took advantage of the changed situation to effect a second landing in the Crimea. This was made by

Soviet Marines rest in a wood near Sebastopol; WW II.

Soviet paratroopers exit from the dorsal hatch of a Tupolev AN-6 bomber during one of the exercises in the early '30s.

Soviet airborne troops capture a German airfield near Leningrad. Although not very effective, Soviet paratroopers showed considerable skill and high courage in WW II.

Soviet paratroopers assembling in the LZ after the biggest paradrop; Kiev maneuvers, 1935.

the 83rd Red Banner Naval Infantry Brigade, the 55th Guards Rifle Division and the 5th Guards Tank Brigade. Under cover of naval gunfire and close air support, the Soviet landing force advanced steadily against the German rearguards, who were covering the withdrawal from the Crimea. Both operations hold some interest in that one of the participants was Admiral S. G. Gorshkov, the father of the modern Soviet Navy.

September 1943 also witnessed an attempt by Soviet airborne forces to disrupt the German withdrawal across the Dniepr river near Kremenchug, the plan being to drop three regiments, the 1st, 3rd and 5th, some 25 miles behind the enemy's lines. The operation was characterised by woeful ineptitude and

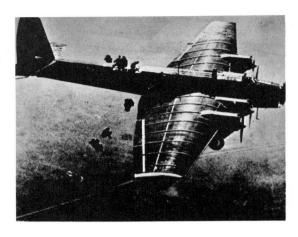

incurred needless loss of aircraft and men during the drop, which also left the regiments widely scattered. Unused to situations demanding a high degree of personal initiative, the paratroops made no effort to regroup and simply dug in where they had landed, being destroyed in detail by the Germans. Hundreds were killed or captured and the survivors were forced to make their way back to their own lines in small groups.

Nonetheless, by 1945 both the Soviet Airborne and the Naval Infantry Corps had learned much, and in the campaign against the Japanese Kwantung Army in Manchuria and its associated coastal operations were to achieve some notable successes. Supported by the guns of the Pacific Fleet and coastal artillery batteries on the Kamchatka peninsula, the Russians landed two rifle regiments on Shumshu Island and others in the Kuriles group. Another force landed on the west coast of South Sakhalin, supporting the ad-

vance of the Soviet 16th Army from the north, while paratroops dropped near Otmari, overwhelming the Japanese garrison after a vicious fight.

Meanwhile, the rapid advance of the Russian armies across Manchuria created ideal conditions in which amphibious and airborne landings could be made to secure important objectives. In this way the Korean ports of Yuki, Rashin and Seishin were captured by Naval Infantry units, while Talien and Lushun (Port Arthur) were taken by surprise airborne attacks, the troops involved being relieved by a fast-moving armored column which had driven the length of the Mukden peninsula. In a very similar operation Pyongyang, now the capital of North Korea,

was secured by paratroops, joined later by an armored column from Wonsan, which had been captured by a naval landing. The crowning achievement of this sequence of events took place on 18 August, when paratroops captured the headquarters of the Kwantung Army and its Chief of Staff, who negotiated the surrender of almost 600,000 men.

By now, the Soviet airborne troops and the Naval Infantry Corps were indeed entitled to regard themselves an elite, and the fullest credit must be given for the imaginative way in which they were handled during the Far East Campaign of 1945, notwithstanding the fact that the Japanese could no longer be regarded as a first class enemy.

Modern Soviet naval infantry in action. Based on their WW II experiences, the Soviets developed an impressive force of elite troops.

33

THE UNITED STATES

The United States Marine Corps has always been regarded as an elite, not simply because of its superb fighting record, but also because of the flexible and imaginative outlook which has enabled it to harness technology to its operations in the form of aircraft, tanks, amtracks and artillery, manned by Marine personnel, all of whom have been through the same gruelling training which produced the Leatherneck rifleman and know just what additional effort is being demanded of them.

During the darkest days of World War II it was the incredible stand of the small Marine garrison on Wake Island that did more to restore bruised American morale than any single event. For sixteen days the Marines held off a large Japanese invasion force, repulsed a landing, shot down aircraft and seriously damaged warships before being forced to surrender. At one stage a faint signal was received from the garrison, specifying its most pressing needs. The signal contained the passage '...send more...Japs....' The two

us Marines go ashore on Makin Atoll.

phrases were obviously not connected, but they were seized upon with delight as being the essence of the Marines' will to win – Send more Japs!

It was in the spirit of Wake Island that the USMC fought throughout the Pacific War, from the bloodbath of Tarawa, through Guadalcanal to Iwo Jima and on to Okinawa. Its victories were gained over an enemy who fought savagely and to the death as a matter of course, and were therefore not cheap. Throughout, the Corps' flexible outlook enabled it to deal not only with the peculiarities of the Japanese method of fighting, but also with the tactical problems caused by landscapes which varied between bare coral atoll and dense jungle, swampland and mountain.

The Corps also produced an elite within an elite in the form of its four Raider battalions, which were grouped together in March 1943 to form the 1st Raider Regiment. The function of these units was to make diversionary raids, to spearhead assault landings by seiz-

ing important tactical features, and to harry the enemy without mercy in appropriate circumstances. An example of the last mentioned occurred in November 1942 on Guadalcanal, when the 2nd Raider Battalion (also known as Carlson's Raiders in honor of its commanding officer, Lt-Colonel Evans Carlson) repeatedly savaged the flanks and rear of the retreating 230th Regiment, one of the best in the enemy's service. Moving parallel with the Japanese column, Carlson's men staged a dozen ambushes, striking hard and then vanishing quickly into the jungle. In this series of actions 500 men of the 230th Regiment were killed, but only 17 of Carlson's Raiders.

Impressed by the performance of the British Commandos, the US Army formed similar units, following in the tradition of the great Robert Rogers by calling them Ranger battalions. The first was raised in Northern Ireland in June 1942 with volunteers drawn from the 1st Armored and 34th Infantry Divisions, being trained by British instructors at the Commando center at Achnacarry in the Scottish Highlands.

In general, the Ranger battalions were employed in a manner similar to that of the USMC's Raiders. The 1st, commanded by Lt-Colonel William Darby, took part in the Allied *Torch* landings in North Africa, capturing the coastal fort of Arzew. It went on to inflict heavy casualties on the Italians holding Sened, to hold the line during the Kasserine Pass debacle, and, using its climbing skills, to unlock the strategically important El Guettar defile with a dashing attack into the rear of the enemy's positions.

General Eisenhower was impressed with the Rangers and at the end of the Tunisian

Marines on Roi Island. The Japanese utilized these smashed palm fronds as cover for their machine gunners and snipers.

us Marine in the Pacific.

LEATHERNECKS USMC

Lieutenant Hawkins peered through the curtain of ever-mounting smoke at the dust-covered beach. The thunder of thousands of shells roaring from the battleships, and of aircraft screaming overhead toward their targets inland, provided opening music for the drama to come.

Hawkins and his men comprised one of the first platoons to assault the Beito Island beaches on 20 November 1943, in the first Allied operation in the Central Pacific – the taking of the Tarawa Archipelago. This chain of naturally fortified islands, fringed by coral reefs, made conventional amphibious assault impossible. The USMC had to rely on LVTs – originally used as amphibious loading trucks to transport the Marines through natural obstacles three miles inland and as close as possible to the enemy.

The 2nd Marine Division (reinforced by the 8th Marine Regiment) approached the beaches in their LTVs, covered by US Navy destroyers. Despite the fire from Japanese shore gun batteries, the LVTs pressed forward in wave attack formation. They hit the beach at 0850 hours; blasting through the Japanese barriers, the Marines soon entered the island lagoon. Now, as they took cover, the real hell of Tarawa began.

With the first and second waves ashore, the following units assembled their landing craft (LC) in the lagoon itself, prevented from advancing closer to the beach by the low tide. The Japanese concentrated their fire on the hapless Marines, but the attackers kept pushing through the inferno, slowly pressing the Japanese inland. More and more Marines equipped with tanks and guns kept pouring in, and by nightfall, the 2nd Marines had a stronghold on Beito, most of which was still held by the Japanese garrison. It took two more days of bloody struggle before the island was finally secured by the Marines on 23 November.

Beito Island had been taken, but at a tremendous cost; 984 Marines were killed and 2,072 wounded. Of the entire 5,000-man Japanese garrison, only 146 survived.

US Marines storm the beach of Iwo Jima.

campaign their strength was increased five-fold. They fought an epic battle against odds at Gela, Sicily, and took part in the bitter fighting around Salerno. In January 1944 they landed at Anzio and were subsequently detailed to spearhead an attack on the town of Cisterna. This ran head-on into a massive enemy counter-attack intended to cut the Allied beach-head in two. The 1st and 3rd Battalions were cut off and fought to the last man, despite repeated and desperate attempts by the 4th Battalion to break through to them. At Anzio the Rangers sustained higher casualties than in all their previous operations put together. Some of the survivors returned to the United States and joined the 5th Battalion, which landed at Omaha Beach on D-Day; others were posted directly to the Canadian-American Special Service Brigade, another elite formation which had been raised for operations which, for one reason or another, presented unusual tactical difficulties.

The US Army's equivalent of the Chindits was the 3000-strong 5307th Composite Unit (Provisional) commanded by Brigadier-General Frank D. Merrill, and better known as Merrill's Marauders. The Marauders penetrated Japanese lines in Burma at the same time as the Chindits, fighting in support of Lt-General Joseph Stilwell's Chinese divisions which were striving to reopen the vital road linking northern Burma with China by means of an offensive down the Hukawng and Mogaung valleys. They operated against the enemy's rear, absorbing a major part of the Japanese reserves in a series of hard fought actions, and led the advance on Myitkyina, where they seized the airfield. Myitkyina town, however, continued to offer

82nd Airborne para-troopers bringing in captured ss soldier.

the most stubborn resistance, but despite the fact that the Marauders were now as decimated by disease and as exhausted as the Chindits, 'Vinegar Joe' Stilwell continued to throw them into action and even had the hospitals combed for those who could stand and carry a rifle, notwithstanding a promise that they would be withdrawn once Myitkyina had been reached. In this way the unit was bled white of its original members, few of whom remained to see it re-designated the 475th Infantry Regiment and in due course awarded colors emblazoned with the battle honor *Myitkyina*. At the time, little was known outside Burma of this scandalous treatment of men who had fought above and beyond the call of duty.

In the field of airborne warfare the US Army developed the largest operational force of any of the combatant nations, yet so many mistakes occurred during the 82nd Airborne Division's drop on Sicily in July 1943 that the future of the entire arm was for a while in serious doubt. Fortunately, an analysis of the operation revealed that the fault lay not with the concept but with the detailed planning and staff work.

The experience was put to good use, for when the 82nd jumped again on D-Day, in company with the 101st Airborne Division, few of the same errors were made. Both divisions were dropped some miles behind the German defenders of Utah Beach, and although they were more scattered than had been anticipated they quickly reorganised themselves and began eliminating opposition in the area, enabling the beach-head to be consolidated.

On 17 September 1944 both divisions dropped on Holland as part of the 1st Allied Airborne Army. The operation, codenamed Market Garden, was intended to seize a bridge across the Rhine at Arnhem with the British 1st Airborne Division, while the 82nd captured the Maas bridge at Grave and the 101st several canal bridges north of Eindhoven, thereby providing a corridor up which the British XXX Corps could advance to the Rhine and, hopefully, beyond. The American divisions succeeded in capturing their objectives after heavy fighting, the 101st being relieved by XXX Corps on the 18th and the 82nd the following day. Together, the 82nd and troops from xxx Corps moved north to capture the bridge at Nijmegen, but further attempts to break through to the now fiercely embattled 1st Division were halted

by determined German counter attacks. Arnhem had indeed proved to be 'a bridge too far', but the 82nd and 101st had proved themselves to be among the most formidable Allied divisions in Europe.

During the Battle of the Bulge the 82nd fought on the northern hard shoulder of the enemy's penetration, while the 101st earned immortality as the defenders of the strategically vital communications centre of Bastogne. When called upon to surrender, Brigadier-General Anthony McAuliffe, the besieged 101st's acting commander, replied with a single word – Nuts! It was the Para's equivalent of 'Send More Japs.'

The last American drop in Europe took place during the Rhine Crossing of 24 March 1945 in which the British 6th and us 17th Airborne Divisions landed behind those enemy units manning the river defenses. It was a model operation, the success of which was never in doubt.

In the Pacific theater of war the us Army also employed airborne troops during the recapture of the Philippines. The major formation was the 11th Airborne Division, which executed a number of well-planned operations, but complained that much more could have been made of its potential had the higher command shown suitable imagination and aggression. The same could hardly be said by the 503rd Independent Parachute

Infantry Regiment, which dropped directly onto the island fortress of Corregidor. In a desperate battle lasting 11 days the 503rd incurred 800 casualties, but 4700 Japanese are known to have died. On the morning of 27 February 1945 the last enemy survivors blew themselves and their arsenal to pieces in a tremendous explosion.

OTHER NATIONS

In general, Italy's armed forces during the period 1940-1943 were poorly motivated, ill-equipped and indifferently led. One brilliant exception was the Navy's 10th Light Flotilla, which developed the Human Torpedo, the frogman, and the Explosive Motor Boat (EMB). This little unit contained many men of imagination, high courage and determination and, for its size, inflicted staggering loss on the Royal Navy. On 26 March 1941 six EMBs penetrated the defenses of Suda Bay, Crete, and sank the cruiser *York* and a merchant ship. A second similar raid on Malta in July ended in disaster, but in September Human Torpedo teams worked their way into Gibraltar, sinking a tanker and three merchant vessels. The 10th Light Flotilla's crowning achievement took place during the night of 18 December 1941 when Human Torpedoes entered Alexandria harbor, crippling the battleships *Valiant* and *Queen Elizabeth,* seriously damaging the destroyer *Jervis* and blowing the stern off a tanker; it was not a bad night's work for six men.

The Japanese made little use of elite forces as we understand them. The Army and Navy, constantly in competition, developed small independent paratroop forces, but used these in support of minor operations to secure tactical objectives such as airfields and oil installations, particularly in the Dutch East Indies. Whether the *Kamikaze* (Divine Wind) movement, involving suicide attacks by aircraft and EMBs against Allied shipping, as well as units formed specially to blow themselves and an enemy tank apart, can ever be classed as an elite save in the spiritual sense, remains highly questionable.

General Frank Merrill, of Merrill's Marauders.

POST WORLD WAR II

None of the forty years since the end of World War II have passed without at least three simultaneous armed conflicts taking place somewhere in the world, yet war between sovereign states has been comparatively rare and, in the main, has been confined to the Middle East. Instead, there have been a number of full scale limited wars, such as Korea and the South Atlantic, which have intentionally been confined to the territory in dispute, and a great many more so-called wars of independence which have varied in their scale between the comparatively low intensity operations of the Malayan Emergency and the high intensity corps-level

In Vietnam, US airborne forces staged massive helicopter operations. The choppers were used in a multitude of roles, mainly for troop transport and evacuation, fire support, and resupply. Here, men of the 60th Inf., 9th Div. dash from a Bell UH-1D helicopter in hopes of encircling a VC battalion south of Saigon.

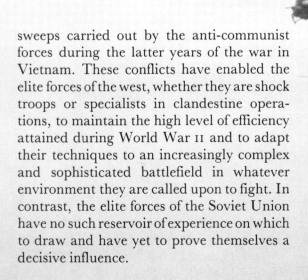

sweeps carried out by the anti-communist forces during the latter years of the war in Vietnam. These conflicts have enabled the elite forces of the west, whether they are shock troops or specialists in clandestine operations, to maintain the high level of efficiency attained during World War II and to adapt their techniques to an increasingly complex and sophisticated battlefield in whatever environment they are called upon to fight. In contrast, the elite forces of the Soviet Union have no such reservoir of experience on which to draw and have yet to prove themselves a decisive influence.

Operation Rangdong –
helicopter assault.

Marine Huey gunship
firing rockets against
enemy positions near
the Marine stronghold
at Marble Mountain,
South Vietnam.

41

THE FAR EAST
VIETNAM 1946-1954

When the French returned to Indo-China in 1946 the majority of Vietnamese made it quite clear that they had no wish to resume the old colonial relationship. Opposition to the French was led by Ho Chi Minh and his Viet Minh party, which embarked on a guerrilla campaign under the direction of Ho's military adviser, Vo Nguyen Giap. For the first two years this was complicated by the civil war in China, which prevented Giap's communist troops from receiving sufficient arms and supplies with which to extend the war, and for much of this period the French managed to contain the threat and even inflicted a sharp check on the enemy at the fortified village of Phu Tong Hoa, in northern Tonkin, during the night of 25 July 1947. The defenses of Phu Tong Hoa consisted of an approximate rectangle with a bastion at each of the four corners and were manned by a 104-strong Legion company of the *3e Regiment Etrangère d'Infanterie*. After a preliminary mortar bombardment which killed both the company's officers, the Viet Minh launched a human wave attack which swamped three of the bastions. The Legion NCOs organised a counter attack which drove the enemy at bayonet point from first one and then another of the captured strongpoints. Thus far the fighting had taken place in almost total darkness, but after moonrise the Legionaires could see to shoot and quickly ejected the Viet Minh from their last foothold inside the defenses, and after a brief but fierce fire-fight across the perimeter wire Giap's men withdrew, leaving their dead littering the area. The company's casualties amounted to 50% (23 killed and 33 wounded), but that did not prevent it from mounting a guard in full ceremonial dress of *kepi blanc,* red epaulettes and blue cummerbund to welcome the commander of the relief column in a manner befitting the Legion's traditions.

In 1949 the communist victory in the Chinese Civil War ensured that Giap's divisions, which had been training hard, would receive ample arms and equipment. However, covering the border with China was the strategically vital Cao Bang ridge, defended by several isolated Foreign Legion posts. In September 1950 Giap concentrated twenty of his best 'regular' battalions on the ridge and on the 16th committed six of these against Dong Khe, which was held by two companies of the 3e REI. The layout of the post was similar to that of Phu Tong Hoa, and during a mass attack on the 17th three of the four strongpoints were lost in heavy fighting which left 40 Legionaires dead and almost 100 wounded. After dusk the remainder counter-attacked and recaptured one of the positions, only to be driven out again; altogether, this strongpoint changed hands no less than eight times during the night before the Legion finally abandoned it. At dawn the Viet Minh closed in for the kill, but were knocked flat or shot down as the surviving Legionaires charged through them in a desperate sortie which emulated the last moments of the Legion's legendary battle at Camerone in Mexico almost ninety years earlier. In this instance, however, the sheer ferocity of the attack enabled the men to reach the surrounding jungle and ultimately to make contact with French troops.

Unfortunately, the dramatic breakout of the Dong Khe survivors was the only comfort the French could glean from Cao Bang ridge. Giap had studied their methods and his forces, present in overwhelming strength, ambushed and severely cut up two relief columns which were moving towards the embattled post, eliminating the equivalent of two Legion battalions. The ridge had to be evacuated and a third battalion, the *1e Bataillon Etrangère Parachutiste*, sustained 90% casualties covering the withdrawal to the Red River; the enemy's loss remains unknown, but their human wave tactics ensured that it was many times that of the Legion.

After the Cao Bang debacle the French were effectively contained within an enclave enclosing Hanoi, Haiphong and the Red

River Delta. During 1951 Giap made several determined attempts to break into this, but on each occasion he was repulsed with heavy loss. For the next two years neither side gained a strategic victory but the French did obtain a number of tactical successes by using their air supremacy in conjunction with parachute troops to defeat the Viet Minh's ambush tactics and establish areas of resistance in the countryside. It was, in fact, the success of this policy which led to the creation of the fortified base at Dien Bien Phu in November 1953. This, it was anticipated, would provoke Giap into fighting the sort of set-piece battle in which he was least comfortable and in which his army would be destroyed by the garrison's firepower and ground attack aircraft operating from safe airfields. The garrison itself consisted of the cream of the French units in Indo-China, the Legion providing three of the seven parachute battalions and four of the ten infantry battalions present, and it was confidently expected that it could be supplied by air.

Both sides knew that the result of the battle would provide a decisive influence on the outcome of the peace talks taking place in Geneva, and both strove hard for victory. Slowly Giap moved four divisions into the area, using an enormous labor force to man-handle his artillery into position on the wooded hills surrounding the Dien Bien Phu basin, together with an impressive array of anti-aircraft guns manned by Chinese regulars. Dien Bien Phu was isolated on 11 January 1954 and the long awaited Viet Minh assault began three months later. Suddenly it was apparent that the French had miscalculated badly – the base lay too far from the Delta for it to be adequately supported or supplied by air, and the enemy's anti-aircraft defense was taking an unexpectedly heavy toll, particularly upon the all-important transport aircraft. Further, ground-attack missions made little impression because of the enemy's excellent camouflage, while approximately one quarter of the air-drop-ped supplies drifted into the Viet Minh's hands, a proportion that was to rise as the perimeter contracted. Simultaneously, Giap's guns and mortars began writing down the garrison's artillery.

The Viet Minh sapped steadily towards each of the base's strongpoints in turn and then swamped them with sheer weight of numbers, regardless of the appalling losses sustained; all too often the fire of the Legionaires was masked by heaps of the enemy's dead or by scores of bodies caught in the wire entanglements. Repeated counterattacks eased the position for short periods, but the final result was always the same. One by one the strongpoints fell and on 8 May the last bunkers were stormed. The French loss amounted to 1500 killed, 4000 wounded and over 8000 captured, half of them Legionaires. On the other hand, the Legion's epic defense of an untenable position had caught the imagination of the world, particularly as it had inflicted a staggering 20,000 casualties on the Viet Minh, the majority of whom were killed.

After Dien Bien Phu the French agreed to evacuate Indo-China and Vietnam was divided into the communist North and the free South. Victory had eluded the Legion but that was the fault of the generals, and it emerged with its reputation enhanced. It had always been employed where the fighting was hardest and during the nine year conflict 11,620 Legionaires had died in Indo-China, a significant proportion of the 75,000 French and Colonial fatalities incurred during the campaign, although it accounted for only 10% of the total French strength. Viet Minh casualties for the same period are estimated at 150,000.

VIETNAM 1964-1973

After the departure of the French the communist administration in Hanoi wasted little time in commencing a guerrilla campaign against South Vietnam, which was allied with the United States. In August 1964 this aggressive policy resulted in a clash between North Vietnamese torpedo craft and American destroyers in the Gulf of Tonkin, followed in February 1965 by a Viet Cong attack on US bases in South Vietnam. From this point a direct American involvement in the war became inevitable, rising to a peak of 542,000 men in January 1969. Of these, however, less than 100,000 were actively engaged in combat at any one time, although this figure includes such elites as the Marine Corps, airborne troops and Special Forces detachments, better known as the Green Berets.

Much had been learned from the French

experience and from the outset American commanders were determined to preserve their freedom of movement, supporting their operations with air strikes, helicopter gunships and artillery located in fortified fire bases. In January 1967 the American Commander-in-Chief, General William C. Westmoreland, embarked on a policy of corps-sized operations which swept large areas of the countryside free of the enemy and thoroughly disrupted his plans. The first of these, Operation *Cedar Falls*, involving two American and one South Vietnamese (ARVN) infantry divisions, the 173rd Airborne Brigade and an armored cavalry regiment, combed out the notorious Iron Triangle north of Saigon, inflicting casualties and capturing 500,000 pages of documents which revealed the entire Vietcong and North Vietnamese (NVA) order of battle as well as the communists' plans. *Cedar Falls* was followed immediately by the even larger Operation *Junction City* and throughout the remainder of the year the communists continued to take serious punishment. It soon became apparent to Giap that he was losing the war and that some spectacular success was needed to restore his prestige and the morale of his forces.

Lying six miles east of the border with Laos and fourteen miles south of the Demilitarized Zone separating the two Vietnams was the Marine combat base at Khe Sanh, the purpose of which was to block communist infiltration through the DMZ to Route 9 and the Ho Chi Minh Trail. Khe Sanh contained a 6000-strong garrison (one battalion 9th Marines, three battalions 26th Marines and the ARVN 37th Ranger Battalion) and there is no doubt that its capture would have profoundly shocked American public opinion. The base was surrounded on 21 January 1968 and for the next 77 days remained under bombardment. Several outlying features were attacked, but while hand-to-hand fighting did occasionally take place, the Marines were using electronic sensors to predict the enemy's movements and were able to inflict heavy casualties with the box barrages fired by their artillery. In addition, a daily average of 300 ground attack sorties, including massive intervention by B-52 bombers, further

discomfitted the communists. Again, although the airstrip remained under constant attack and several aircraft were lost to ground fire, the Americans devised a number of ingenious ways of dropping ammunition and food into the base and the supply situation never became critical. On 6 April Khe Sanh was relieved by Operation *Pegasus,* spearheaded by the crack 1st Air Cavalry Division, and the siege was over.

Tempting as the defeat of an American elite might have seemed to Giap, it has been suggested that he never intended to turn Khe Sanh into a second Dien Bien Phu, and that

South Vietnamese Strike Force 321 fires a .30 cal. machine gun at VC positions in 'D' zone, at Ben Cat in the 'Iron Triangle'; November 1964.

his real objective was to tie down US troops while the simultaneous Tet offensive was in progress. In fact there were never less than two, and sometimes four, NVA divisions present in the area, and this confirms that it was Giap who chose to tie down his regulars while the Viet Cong strove for control of the towns and cities. Yet somehow, careless as he was with his men's lives, he could not steel himself for a final, outright assault; the carnage inflicted by the weary Legionaires at Dien Bien Phu could hardly have been forgotten, and to have exposed his precious regular formations to the ample firepower of the equally tough and quite unshaken Marines was to invite their destruction.

The Tet offensive failed utterly and the Viet Cong was virtually destroyed, losing between 30/40,000 killed, or 80% of its effective strength; the most protracted fighting took place in the former capital of Hue, which was recaptured by Marine, Army and ARVN units. However, thanks to a callow and somewhat frightened press corps which lacked the historical perspective to see Khe Sanh and Tet as desperate attempts by the communists to regain the initiative, public opinion in the United States received a tremendous shock and the anti-war lobby prospered.

either as part of major operations involving regular formations, or in the local context, ambushing communist supply routes and carrying out covert surveillance tasks. Naturally, the communists reacted strongly and mounted regular attacks on SF base camps. More often than not they were beaten off with heavy loss, but occasionally a camp would be overrun, as was that of Lang Vei, near Khe Sanh, on 7 February 1968. Such importance did the NVA attach to this operation that its attack was led by thirteen PT-76 light tanks. Three were destroyed and others

US Special Forces from A Team 215 and Vietnamese CIDG troops on a lunch break. The Americans soon got used to the local food and habits.

Meanwhile the Green Berets, themselves trained specialists in guerrilla warfare, had been painstakingly building up their Civilian Irregular Defense Group (CIDG) program. Small SF teams were inserted into remote rural areas in which the villagers were being forced to supply the communists at gunpoint. Their first task was to arm the people and teach them how to defend themselves, and this proved so successful that in due course CIDG units began operating offensively,

damaged by close range anti-tank weapons and at least three more were knocked out by air strikes and artillery fire from Khe Sanh, but the remainder broke through the perimeter. The garrison, consisting of 24 Green Berets and several hundred CIDG personnel, fought on throughout the day and after dark the survivors, 13 Green Berets and 60 Montagnard tribesmen, broke out and managed to reach Khe Sanh. On 3 March 1969 the NVA again used PT-76 light tanks, this time accompanied by BTR 50 APCs, in an attempt to storm the Special Forces camp at Ben Het in the Central Highlands. Unfortunately for the communists a platoon of M48A3 Pattons was also present in the compound, and after two light tanks and an APC had been blown apart they scuttled off into the darkness.

After 1968 the NVA was placed on the strategic defensive and although never eliminated was unable to make headway, despite the steady reduction of American forces in Vietnam and the 'Vietnamisation' of the anti-communist war effort. In 1972 Giap abandoned guerrilla warfare and used his divisions to mount a conventional offensive. The failure of this, coupled with increased American air activity over North Vietnam and the complete collapse of the communist air defence system, compelled Hanoi to call an end to hostilities. The communists, however, had no intention of keeping the peace and once the Americans and their allies had left they wasted no time in launching a massive invasion of South Vietnam which the weakened ARVN was incapable of halting.

Mobile Strike Force of US 5th SFG and Vietnamese troops cross a river during a two day search and destroy patrol. At this river crossing, the force drew sniper fire from VC positions on the far side.

SPECIAL WARFARE
VIETNAM 1961-1975

The conflict in Southeast Asia was a classic example of Unconventional Warfare on all sides. Among the various US units to carry out UW were the Special Forces, Navy SEALs, and several Military Intelligence groups. One of the earliest instances of SF involvement in South Vietnam was that of the Mobile Training Team (MTT). Deployed from the 1st SF Group (Okinawa), the MTT began training South Vietnamese Army (ARVN) troops in counter-insurgency operations. With a sensitive approach to the country's multi-ethnic population, the 'Green Berets' won the proverbial hearts and minds of the trainees, many of whose tribes and people had previously been neutral or associated with the VC.

THE GREEN BERETS JOIN IN THE FRAY

The Special Forces started to fight in Vietnam even before the war became publicized. They entered their second decade of existence in a rapid buildup to what then was seen as an inevitable war. In 1961, the 5th SF Group was activated, following the 10th (1952), the 7th (1953), and the 1st (1957). In 1963, the first detachments of SF units were sent to Vietnam to render military assistance. The US Army Special Forces/Vietnam was formed in Saigon in September 1962 (moving to Nha Trang in 1963), with personnel from the 1st SFG and elements of the 5th and 7th Groups. Its mission was to advise and assist the South Vietnamese in the organisation, training, and employment of the Civilian Irregular Defense Group (CIDG). The first training camp was established near the village of Ban Me Thout in 1961. (Later, the 5th SFG would be responsible for CIDGs and their missions, including those previously carried out by the CIA, such as border surveillance.) The network of fortified, strategically located camps, each with an airstrip, later proved to be invaluable recce and fire support bases for remote areas.

In order to expand SF activities in Vietnam, the 5th SFG was deployed from Fort Bragg to Nha Trang in October 1964. Its primary task involved 'conventionalization' of the CIDG, and its operations with those of the regular army. New bases were established mainly in border areas, and more men trained. Purposely located astride the major enemy supply and infiltration routes, the new camps were the main base of operations for the group's CIDG personnel. As a result, the bases provided excellent intelligence and surveillance sources, with the patrols operating from them inhibiting enemy movement. Upon contact, these patrols would call in air strikes – A-1Es, Cobras, or bombers deployed nearby. As a consequence, the enemy sought to destroy these camps; the results were usually successful for the defenders, but not without cost. While the North Vietnamese regular army attacks were usually repelled with massive air and artillery support, the enemy reverted to infiltrating the camps, terrorizing them from the inside with CIDG members they recruited. In an attack on the CIDG camp at Nam Dong, SF Capt. Roger H. C. Donlon won the first U.S. Medal of Honor since the Korean war, for gallantry during a savage five hour battle. The 'A' Detachments (see chapter 2) at Nam Dong received a total of 33 US and Vietnamese decorations for their stand against a superior NVA and VC force.

By mid 1965, the 5th SFG had assumed wide-ranging tasks. These included civil assistance, such as dredging canals in enemy-controlled villages. Their ability to wage aggressive combat was finely balanced with their ability to conduct civic actions, such as the building of schools and hospitals.

EYE FOR AN EYE

At its peak in 1966, the 2600 Green Berets of the Special Forces controlled over 80 camps and some 60,000 CIDG soldiers. In later stages of the war, with the training of Vietnamese forces in special warfare and the introduction of mobile strike unit concepts, more Vietnamese were recruited, and long range, cross border patrols commenced. Reaction and exploitation techniques of several continuous missions (Delta, Sigma, and Omega, of which most details are classified even today) carried the guerrilla war to the enemy. This bold move, of fighting guerrilla forces with guerrilla warfare, was abandoned after a while, not due to lack of success, but rather to a lack of enthusiasm by the regular army and unfavorable press reports on 'Special Forces-controlled mercenaries'.

First called the 'Multipurpose Reaction Force', it was made up of four battalions (called Mike Forces) of four companies, each with some 150-200 men, as well as a 135-man recce company. Headquarters and service company (227 men) was selected from CIDG. The force would be assigned an area in enemy-held territory or beyond US forces authority of operations, and deploy either in assistance to CIDG camps, relieving them from enemy pressure, or in sustained guerrilla warfare aimed at the enemy rear. Using many techniques 'borrowed' from the Vietcong, the Mike Forces were highly effective in bringing the war to enemy bases, logistical centers, and lines of communications. The Mikes were also ca-

pable of operating as a brigade. While at the beginning, Mike Force capabilities were not totally appreciated, their missions were later fully exploited, and enhanced with the creation of the 'Mobile Guerrilla Force', made up to 150 Mike Force personnel and commanded by an SF A Detachment. These specially trained Mikes performed operations in many remote zones, previously considered VC/ NVA 'safe areas', during 30-60 day sorties called 'Blackjacks'.

A typical MGF Blackjack started with the Mikes being clandestinely inserted into an area of operation. While all Mike Force soldiers, CIDG and US, were airborne-qualified, only a few Blackjack missions were inserted by parachute. Most employed 'Slick Hueys' – helicopters using deception in landing their passengers. During Blackjacks, these troops were normally out of contact with their bases for periods of up to a month. Resupply was also done clandestinely. One of

5th SFG team firing an 81mm mortar; Nha Trang, February 1967.

the common procedures employed specially equipped Vietnamese Skyraiders carrying napalm canisters filled with food, clothes, and ammunition; these were dropped at predesignated spots on the Blackjack route and immediately collected. The Skyraider would then hit a nearby spot with real napalm, completing the deception. The valuable intelligence gathered by these units was indispensable. In 1967, all Mikes and MGF were integrated into the 'Mobile Strike Force' (MSF) which operated successfully in South Vietnam until 1970.

On top of their UW roles, Special Forces, CIDG, and MSF personnel were effective in intelligence gathering through 1971, when the last SF unit was sent home from Vietnam. They supplied about half the information acquired by the Military Assistance Command/Vietnam (MACV). In 1968, the CIDG also proved its worth in countering the NVA

SF; Thailand.

Tet Offensive, and in 1970, during operations in Cambodia, it paved the way for missions into the enemy mountain and forest strongholds.

'Vietnamization' was practiced by the Green Berets since 1961 when they began their work in the country. Instead of fighting for the Vietnamese, they taught the soldiers and civilians how to defend their homes. During almost a decade of SF contact with the Vietnamese, the Americans worked with many ethnic groups, including the Montagnard tribes, Khmer Cambodians, Hoa Hoa and Cao Dai, Numg, Rhade, Thai, Lao and Kha & Meo minorities in Vietnam, sharing their lives with the natives, and advising and assisting in the defense of their villages while at the same time fulfilling their mission.

49

PLAYING DIRTY

Deep in-country and far from the eyes of the press, other UW operations were carried out by the MACV's 'Studies and Observations Group' (cover name for Special Operations Group) assisted by South Vietnamese units. These SOG raiders directed the MSF in engaging enemy units and bases, as well as in hitting targets of opportunity or calling in air strikes. Most of their operations, however, were independent in nature.

In 1958, the South Vietnamese government created its secret service, commanded directly by President Diem. It was redesignated as the Vietnamese Special Forces Command after the 1963 overthrow of Diem. Supported and financed by the CIA, command was assumed by MACV-SOG in 1964. Originally headquartered in Cholon, MACV-SOG transferred to Saigon in 1966, with its 'Air Studies Group' based at Nha Trang, and its 'Maritime Studies Group' at Da Nang together with its Forward Operations Base 1. Launch sites were based at several border villages and at CIDG camps. A 'Psychological Studies Group' (Psyops) was located in Saigon, with up to 70 transmitters and remote antenna stations used for Psychological Warfare against North Vietnam. At its peak, MACV-SOG was assigned 2000 Americans, most of them Special Forces, with over 8000 highly trained Vietnamese troops. The mainstay of its autonomous 'air force' was the 90th Special Operations Wing with UH-1F 'Green Hornet' helicopters heavily armed with guns and rockets, a squadron of C-130 Hercules, C-123s used for covert operations (piloted by Chinese Nationalists) and H-34s from the 219th Vietnamese AF Squadron. The Naval forces included US SEALs and Vietnamese UDTs. The ground forces of MACV-SOG were organized into Southern, Central, and Northern commands.

These groups operated in utmost secrecy with only a handful of commanders even knowing of their existence. Augmented by some 940 US SF troops arriving in 1966 under the cover designation 'Special Operations Augmentation', special operations expanded well beyond the borders of South Vietnam. MACV-SOG was also eventually responsible for several forces operating with South Vietnamese Rangers, such as Detachment B-53, which was involved in highly classified operations.

Bread and butter for these forces were the various cross-border operations regularly conducted against VC, Khmer Rouge, Pathet Lao, and NVA elements in their own territories; tasks such as kidnapping, assassination, and the inserting of rigged mortar rounds into enemy ammunition supplies (which were set to explode while being handled by the crew) were usual. Another invaluable (but all-too-often unrewarding) effort was to keep track of US POWs, conducting raids and assisting in escapes from accessible camps. Training and dispatching agents into North Vietnam was another effort, and various psychological warfare operations concluded the list. In several cases, SOGs were sent after lost or captured documents, which were as hot as explosives....

SOG command and control had up to 30 Spike Recon Teams (RT), each with 12 men, 3 of them SF, the rest excellent jungle fighters hand-picked from CIDG. The RT could set long range ambushes, call in air strikes, and blow up ammunition caches; they were the primary recce element and source of field information available to SOG. Another element was the Hatchet Force, composed of 5 SF and 30 indigenous personnel, acting primarily as a fast RT reinforcement. There were also four Search-Location-Annihilation Mission companies (SLAM) which were assigned to exploit 'promising situations' and were also capable of independent operations. Cross-border forays began in the north as early as 1964, and into Laos in 1965 with operation 'Shining Brass', an effort that continued through 1971 (renamed 'Prairie Fire') when it was turned over to CIA-supported Laotian tribesmen commanded by Gen. Vang Pao. 'Shining Brass' saw 12 Spikes inserted into Laos as well as North Vietnam, primarily to locate potential aerial strike targets. Operations in Cambodia began in 1967. Before these cross-border operations, the main field of deployment was limited to VC-controlled areas in South Vietnam. Among the largest in scale and diversification was Operation 'Leaping Lena', aimed at tracking and monitoring the Ho Chi Minh trail inside Vietnam with multiple six-man teams.

THE GREEKS

With the SOG becoming more and more involved in active operations inside Vietnam, a dedicated force, designated 'Project Delta', took over 'Leaping Lena'.

Project Delta was the prototype of a specialized force, later followed by similar units also designated by Greek letters. Delta consisted of a single A Team (SOG) which led CIDG and VNSF in long-range reconnaissance patrols. By June 1965, the 5th SFG assumed a more active role in the project, and assigned Detachment B-52 for command and control of Project Delta. By this time, with MACV's growing dependence on its information, Delta assumed the responsibility for intelligence (both tracking and post-attack assessment), hunter/killer missions, special purpose raids, and harassment and deception strikes. Project Delta grew into six, later 12 CIDG 'Roadrunner' teams (for long distance enemy trail recce), a Nung Camp security company, and a

US Army UH-1D Hueys transporting troops in Vietnam. Note the helmeted side door gunner.

South Vietnamese Ranger Battalion (91st Airborne) as a reaction reinforcement unit. Delta was also involved in training US troops in long range patrol tactics. Although it was based in Nha Trang, it was capable of operating freely in all areas, directed by MACV and South Vietnam JGC.

With the growing importance of Project Delta and the increasing demand for its services, Detachment B-52's capabilities were taxed to the limit. So in 1966, another unit, Detachment B-56, was authorized to head the new 'Project Sigma'. It contained eight recce teams, three commando and one camp defense company. Operating out of Ho Ngoc Tau from 1967, B-56 used Cambodian and Chinese CIDG groups for its missions.

'Project Omega' was handled by detachment B-50, and was involved mainly in special recon missions in response to the growing need for intelligence. Project Omega performed long range reconnaissance missions beyond those furnished by Project Delta. It was composed of four (later eight) Roadrunner teams, and eight (later 16) reconnaissance units conducting saturation patrols throughout specific zones. Backing up these elements were three commando companies used to exploit small unit contacts, aid in the extraction of compromised teams, and perform 'reconnaissance-in-force' missions. Omega also had a camp defense company. Operating out of Ban Me Thout and a number of CIDG camps, Omega was mainly operational within the I Corps area.

Then there was 'Project Gamma', manned by Detachment B-57 as an information gathering element aimed mainly against NVA infiltration into Cambodia. In 1968, Project Gamma was operating from camps in II, III, and IV Corps areas, supplying timely and valuable information on enemy operations across the border.

THE SEAL (Sea Air Landing) TEAMS IN VIETNAM

With the increasing involvement of the US in Vietnam in the mid '60s, it was felt that Navy SEAL platoons would be readily adaptable to operate in river areas against VC and NVA sappers. Starting out in 1966, results became sufficiently impressive to warrant the deployment of additional units to the country. The two platoons from SEAL Team One (west coast) already on station carried out ambushes, and manned listening posts in areas well back from the main river where the VC regularly operated. SEAL Team Two (east coast) then took charge and considerably expanded operations and involvement in the conflict. On these missions, the SEAL elements would usually use small craft which were heavily armored. As the range of operations grew, longer transits and greater demands for high speed and firepower made it necessary to develop a new type of craft. Now, with improved performance vehicles (which were also much quieter then the previous ones) SEAL casualties were much reduced.

SEAL forces in 1966 were composed of the SEAL platoons, closely coordinated with Boat Support Units (BSU), Mobile Support Teams (MSU), and Seawolf-Huey helicopter air patrols. This combined force was later designated the 'SEAL Package'. The officer in charge could therefore conduct much more effective operations, getting timely support with less effort than 'through the channels'.

After starting out with some difficulty in the new and rather disorganized arena, the SEALs had become highly praised for their operations and soon led other US forces inland on search missions. With word of their accomplishments spreading fast, SEAL units were assigned more missions in 1967. Their manpower increased and area of responsibility grew larger.

SEAL squads also operated in offensive missions against VC installations. In 'Operation Charleston', they blew up VC wells, acting on intelligence they themselves gathered the previous month. In March 1968, a communist defector led the SEALs to a VC gun factory which they later destroyed along with two arms caches. In 1970, the SEALs made a successful attempt at POW liberation when a team of 34 men (led by 15 SEALs) broke into a prison camp, fought the guards, and released all 22 South Vietnamese POWs. The SEALs were also involved in long-duration intelligence gathering, manning listening posts on waterways for up to seven days at a time, checking the VC movements. With the introduction of new, high-speed 'Boston Whaler' speed boats in 1967, quick recovery became available for SEAL squads ambushed or caught under fire.

Switching to mobile operations, the SEALs conducted armed Provincial Recce Units (PRU) which, similar to the 'Greeks', undertook mobile operations aimed at the VC hinterland. The PRUs combined both US SEALs and indigenous personnel. These PRUs were tasked mainly with intelligence gathering on key Vietcong, and, acting as hit squads, were ordered to 'neutralize' them whenever possible.

US Special Forces in Southeast Asia. A weapons specialist aims a 3.5″ rocket during an orientation course in Unconventional Warfare; Thailand, 1967.

Another member of the A Team is the chief medic, here during a civic aid action program run together with the military assistance efforts in Thailand.

A formation of AH-IG
Cobra gunships over
Vietnam; 1970.

Winged assault in
Vietnam; Huey UH-ID
helicopters landing
forces in a jungle area.
They are armed with
side-mounted M60
machine guns.

THE MALAYAN EMERGENCY 1948-1960

As a result of an international communist decision in 1948 to spread the so-called 'armed struggle for freedom' throughout the former colonial territories in the Far East, the Malayan Communist Party formed a military wing known as the Malayan Races Liberation Army and immediately commenced a guerrilla campaign of terror and intimidation against the civil authorities and the security forces. From the outset, however, this campaign suffered from two fundamental weaknesses. The first was that Malaya had already been promised independence, and the second was that despite its title the MRLA was 90% Chinese in origin. Nonetheless, despite failing to gain popular support, the communists continued with their murderous attacks and assassinations.

The British tackled the problem simultaneously on several levels. First, the terrorists' influence among the rural population, and especially among the Chinese community, was reduced by providing the protection of fortified villages. Secondly, an efficient intelligence apparatus was set up which gave the security forces an insight into the terrorists' local aims and objectives. Thirdly, once contact was obtained with a terrorist group in the jungle the security forces embarked on a relentless pursuit which did not end until the group had been destroyed or dispersed.

It was a war of sustained effort and small clashes which enabled every section of the British Army and the Royal Marines to perfect both its jungle warfare and its counter-insurgency techniques, and in particular it led to the Special Air Service Regiment, which had been disbanded after World War II, being re-formed in 1950. In Malaya the SAS was significantly influenced by the Chindit penetrations in Burma and undertook long-endurance operations in the jungle. As always, it adapted its techniques to its surroundings. Thus, while others might think twice about parachuting into rain forests, the SAS perfected the Tree Jump, each man carrying a long rope with which to let himself down when his parachute snagged among the high branches. It also chose weapons which were eminently suitable for close quarter combat in the limited visibility of close bush, notably the pump-action shotgun, which could be handled with speed and deadly accuracy.

By the time the last terrorist gangs had been eliminated in 1960 over 3000 civilians had been murdered and almost 2000 members of the security forces had been killed. The MRLA, however, had sustained 6711 killed, 1289 captured, and 2704 of its members had given themselves up; further, the communist dream of controlling Malaya's strategic assets of tin and rubber was now a dead duck.

KOREA

The communist invasion of South Korea began on 25 June 1950 and for a while swept all before it, pinning the United Nations' troops inside a perimeter around the port of Pusan in the south-eastern corner of the country. On 15 September the position was dramatically reversed by an amphibious landing at Inchon on the west coast, executed by the 1st Division USMC and the 7th US Infantry Division. Within a week Seoul had been captured and the North Koreans, in danger of being cut off, were fleeing north as fast as they could travel.

The Inchon landing, a strategic master stroke, presented serious technical problems because of strong currents and tidal rises, but the expertise inherited by the Marines from their vast experience during World War II enabled them to solve these without undue difficulty. As part of the deception plan a

small party of Royal Marine swimmers carried out a simulated beach reconnaissance at Kongso-on, 14 miles south of Inchon, while a US Army Raider company staged a noisy demonstration on Robb Island offshore, concentrating the North Koreans' attention on that sector of the coast.

After Inchon and the breakout from Pusan the 1st Marine Division, now part of the United Nations' X Corps, was shipped to Wonsan on the east coast and took part in the general advance to the Yalu River. By now the North Koreans had been utterly routed, but on 25 November Chinese divisions swarmed across the border to their aid. The Eighth Army, as the UN forces in Korea were now known, was again forced onto the defensive and X Corps was compelled to retreat on

the port of Hungnan, from whence it was evacuated to Pusan. This fighting withdrawal, in which the 1st Marine Division was joined by 41 (Independent) Commando, Royal Marines, was made in sub-arctic conditions and remains one of the most remarkable feats in USMC and Royal Marine history.

After reorganising at Pusan the 1st Marine Division moved back into the line in January 1953 and played a major part in containing the renewed Chinese offensive. By now, the enemy's human wave assaults were running into the Meatgrinder defense devised by General Matthew Ridgway, the UN field commander, and the carnage inflicted was such that not even the Chinese Red Army could absorb it. Gradually the Eighth Army went over to the attack and by March the

The Chindits in Burma, during WW II. The SAS used the Chindit's experiences in their Malayan operations.

enemy was retreating all along the line, I Corps dropping the 187th Airborne Regimental Combat Team on Munsan-ni, 20 miles north-west of Seoul, and then pushing out an armored task force to join it.

At the end of April the Chinese, reinforced, attacked again. In the center of the line the Marines' flank was exposed when a neighboring South Korean division gave way, but an orderly withdrawal was carried out to fresh positions on the Hongchon River, where the line was again stabilized. By the third week of May the enemy had fed all his available manpower resources into the Meatgrinder, had out-run his supplies, and was on the point of disintegrating. Once more the Eighth Army took up the pursuit and by 24 May the 187th Airborne RCT and the 1st Marine Division had reached the Hwachon Reservoir and Inje. For political reasons the advance did not continue far beyond and the front was consolidated on the strongest possible line.

Meanwhile, 41 Commando had been making life miserable for the enemy along his eastern seaboard, along which ran the strategically important railway carrying military supplies from China and the Soviet Union. Landing from American submarines and warships, Commando parties carried out regular raids on this, destroying trains, track, culverts, embankments, bridges and tunnels. This not only disrupted the enemy's already inadequate supply line but also forced him to tie down large numbers of troops which he could ill spare to defend the line along its entire length. The Commandos also established a permanent presence on the islands of Wonsan harbor, which were used as a forward base for clandestine missions, and provided accurate spotting for naval bombardments. By the time 41 (Independent) Commando was withdrawn from Korea in December 1951 its members had received 15 British and 14 American awards for gallantry.

On 23 June 1951 the Soviet Ambassador to the United Nations proposed a ceasefire, and although positional warfare was to continue for a further two years the communists clearly understood that a military victory lay beyond their grasp.

THE BORNEO CONFRONTATION 1963-1966

In August 1962 the three northern states in Borneo (Brunei, North Borneo and Sarawak) joined the Federation of Malay States to form one economic and political unit henceforth known as Malaysia. This did not suit President Achmed Sukarno of neighboring Indonesia, whose ambitions not only extended to the whole of Borneo, but also to the Malay peninsula as well, and he publicly announced that he intended to crush the new country. As the United Kingdom was committed to the military support of Malaysia an armed confrontation became inevitable.

The fighting was confined to the island of Borneo and its coastal waters. After attempts to provoke insurrections in the Malaysian towns had been easily crushed, Indonesian regular forces embarked on a policy of infiltration and raiding across the long common frontier. The joint British/Malaysian strategy was based on control of the rivers, which were often the only highways available, and domination of sensitive border areas.

The British commitment included, among others, such elites as the 3rd Commando Brigade and its Special Boat sections, SAS squadrons and Gurkha battalions, all thoroughly schooled in the techniques of jungle warfare by the long Malayan Emergency. The Indonesians also employed some of their best troops, but these were no match for their opponents and were repeatedly repulsed in a series of small but nonetheless bloody clashes. Having sustained the loss of 590 killed, 222 wounded and 771 captured, Sukarno agreed to a ceasefire in 1966, having accomplished nothing. Commonwealth casualties during the confrontation amounted to 150 killed and 234 wounded.

MIDDLE EAST

WAR IN THE SHADOWS
ISRAEL 1948-1982

World War II was in its fifth year. There was scarcely a nation in Europe that had not felt the iron might of the Axis powers. Now one more – Hungary – was about to join the ranks of the conquered. To many, the peril facing that small country was insignificant compared to the havoc Hitler had already wrought. But to the Jews of Palestine, the fall of Hungary would mean another million of their brethren exposed to the horrors of the Nazi regime, a power that had already wiped out millions of Jews all over Europe.

Time and time again, the Jewish authorities in Palestine importuned the British to let them help the Jews of Eastern Europe. The Haganah had worked out an ingenious, though exceedingly daring, plan: a team of Jewish volunteers, born in the countries now

under siege, would train as parachutists. Their training completed, they would be dropped into Eastern Europe by the RAF; they would then join the local partisans and do whatever they could to sabotage the Nazis. But His Majesty's Government was not easily convinced. To the British, this idea seemed more like a suicide mission dreamed up by a frustrated people than a workable plan whose chances of success could justify the risk – and the expense.

Finally, the British military authorities at Cairo GHQ offered the Jews a deal. A limited number of hand-picked Jewish volunteers would be accepted for parachute training at an RAF facility. If they passed the gruelling course, they would be dropped into Eastern Europe. However, they would be under strict

The last photo of the group, taken by a British officer; Anske-Bistortza airstrip, Slovakia. At the extreme left is Yoel Palgi, the only survivor of the group. In the background is a converted B-17 used for clandestine operations.

orders regarding their mission. Their main responsibility would not be to rescue Jews, but to set up a network of underground operatives who would offer aid and assistance to Allied airmen forced to bail out over hostile territory. The agents would help them to evade capture and to return to their own countries. This was not exactly what the Haganah had had in mind. Still, it was better than the obstinate refusals they'd been getting, and they decided to accept the offer.

Immediately following the Haganah's acceptance of the British terms, a call went out for volunteers. Hundreds of Rumanian and Hungarian-born Jews responded, but only a handful were chosen. These included two girls, Hanna Szenes and Haviva Reik – not surprising in view of the fact that all Jewish paramilitary organizations accepted women into their ranks. What was amazing – at least to the RAF instructors – was the dedication and tenacity the Jews showed during the arduous training at the Middle East Parachute Training Center at Ramat David, not far from the kibbutzim from where most of the volunteers had come. Despite their total lack of military background, they had a sense of purpose which carried them through the hardest stages of the course. Remarkably, the British instructors noted, Hanna Szenes

and Haviva Reik led their class at the end of the training period, having achieved better all-round performance than any of the men! Finally, the course ended. The Jewish volunteers, dressed in flight suits over RAF officers' uniforms, were flown out of Palestine to an airfield somewhere in Italy. There they were loaded into the black-camouflaged Halifax which was to take them one step closer to their personal, if unofficial, goal: the rescue of Hungary's Jews.

Friendly hands reached out of the darkness. Partisans, apprised of the Haganah parachutists' impending arrival, had arrived to guide them through the forest and across the border into Hungary. Moving stealthily in the night, the Jews followed their guides – straight into the waiting arms of the Budapest Gestapo. Hanna Szenes and her party were tortured for days on end, but nothing the Nazis could do was capable of making them betray their cause. Before her execution by firing squad, the quiet, brave kibbutz girl

Three of the 'Hungarian' party in disguise as tramps for the drop into Hungary; 1944.

A Special Night Squad listening post on the Lebanese border.

wrote a short poem. Every Israeli para-trooper knows it by heart.

'Blessed is the match consumed in kindling flame.
Blessed is the flame that burns in the secret fastness of the heart.
Blessed is the heart with strength to stop its beating for honor's sake.
Blessed is the match consumed in kindling flame.'

The story of Hanna Szenes is a saga in itself, an inspiration to freedom fighters throughout the world. Another member of Hanna's team, however, was to have even more bearing on Israel's military history. Sentenced to a concentration camp, Yoel Palgi managed to jump from the camp-bound transport train and eventually make his way back to Palestine. Four years later, he was to be given the task of founding a Parachute Corps for a fledgling army called the Israel Defense Forces.

51st Commando: The Jews Fight Fire with Fire

If the first step toward an IDF Parachute Corps was taken by Rumanian and Hunga-

rian Jews, the first Jewish commandos since Bar-Kochba's rebellion against Rome were born in Germany. Many young German-born Haganah members, inspired by the British commando units established in an effort to check Rommel's advance into the Middle East, appealed to the British Army to let them form a unit of their own. Knowing the language and customs well, they were ideally suited for undercover work both in Germany itself and behind the German lines in Europe and Africa. In addition, their unmilitary appearance was perfect for commando work. The British, however, were reluctant; they seemed to feel they had enough on their hands without the apparent contradiction in terms of a Jewish commando force.

As the war continued and Rommel's forces pressed on unchecked, the impending danger drove the British to reconsider. Grudgingly, they agreed to establish a small 'German unit' – German-born Jews trained in Wehrmacht drill and tactics by two defectors from the German army. This highly secret unit was trained, of all places, in a remote corner of a kibbutz in Palestine! Those kibbutzniks who chanced to run across a squad of stiff-spined soldaten goose-stepping behind the cowbarns to a cadence of 'Links! Rechts! Links!' would smile grimly and pass on, aware of just how important – and how secret – this band of masqueraders was. Finally, their training concluded, the Jews of His Majesty's 51st Commando were transferred to a secret base in Egypt, ready to strike behind the German lines.

The first mission of the newly-created commando team, headed by a German-born British Army captain, was to blow up a German petrol dump near Tobruk. The group set out in captured Afrika Korps vehicles and reached their objective in good order. As they entered the camp, however, disaster struck. One of the Wehrmacht defectors, apparently losing his nerve, gave the show away, and within seconds the 51st Commando found itself under heavy fire. The CO was killed outright; the rest of the group scattered. Some were captured; others, after wandering for days in the waterless Libyan Desert, were picked up by a far-ranging LRDG patrol and

Special Night Squad units searching an Arab suspect. This first type of offensive warfare was conducted against the terrorists in the '30s.

brought back to the British lines. Despite the courage of the German Jews, this setback impelled the British authorities to disband the 51st Commando. Some of its members, however, were accepted into other British commando units, participating in many famous raids. One of them later became a high-ranking officer of the Israel Defense Forces.

The Palmach and Other Jewish Underground Fighting Units

Meanwhile, the situation on the home front was deteriorating. The Haganah realised that His Majesty's Government had a tendency to act in its own best interests. They knew the British would prefer to retreat eastward, should Rommel succeed in breaking through the El Alamein line, rather than stand and fight for the Jews of Palestine. Because of this, they created the Palmach, a regular organization of highly trained fighters, which would be responsible for defending the land and people should all else fail. A top secret contingency plan was even made to concentrate the entire Jewish population of Palestine on Mount Carmel in a last desperate attempt to hold off the German invaders – as the forces of Masada had done nearly two thousand years before. Fortunately, General Montgomery's 8th Army succeeded in driving the Afrika Korps out of Africa.

Palestinian Jewry's delight at the British victory turned to dismay as postwar policy began to evolve. Astonishingly, the British had begun to befriend the Arabs, who had actually sided against them during the fighting! The British lost no time in declaring the Palmach illegal; severe restrictions were placed on all Jewish paramilitary organisations, and frequent house-to-house searches for arms and ammunition were conducted. Worst of all, the White Paper – a British policy statement handed down in 1939, severely limiting Jewish immigration to Palestine – was being rigidly enforced. Survivors of the Holocaust, arriving in decrepit ships hastily acquired by the Haganah in Europe, were being turned back to sea or diverted to detention camps on Cyprus. As World War II drew to a close, the Haganah and the other Jewish paramilitary groups, the IZL (National Military Organization) and the LEHI (Israel

Freedom Fighters), began to train crack clandestine commando units. These represented the decision by Palestinian Jews to save the last remnants of Eastern European Jewry – by any means.

The years between World War II and Israel's independence featured many lightning commando operations designed to frustrate British efforts to keep the Jews out of Palestine. Any British installation involved in those efforts – radar stations, airfields, coast guard patrols – was considered fair game. In a series of extremely daring raids, the underground commandos struck time and again right under the noses of His Majesty's troops. In one particularly spectacular operation, a group of LEHI fighters blew up a whole squadron of RAF Spitfires at Kfar Sir-

Special Night Squad section patrols the Jordan Valley; 1938.

kin, an airfield in the southern Sharon Valley.

Although the many desperate missions of the underground commando fighters did little to change British policy, they did succeed in getting many thousands of Holocaust survivors into Palestine. These successes can be attributed both to the fearlessness and finesse of the commandos themselves, and to the expert military leadership which was developing among the Palestinian Jews. Both were shortly to be put to the test in Israel's War of Independence.

RAID ON KALKILYA – 1956

Raiders capture documents in the Jordanian police station at Kalkilya.

The raiding party at the target; leading is Motta Gur, later the brigade commander and IDF Chief of Staff.

The force assembles at the target. Note the familiar shape of the British-built police station.

A demolition charge is put in place to blow up the target.

Kalkilya police station being blown up, as seen from the forward HQ in Israel. The lights mark the Israeli village closest to the Arab town, only few hundred meters away.

After the explosion.

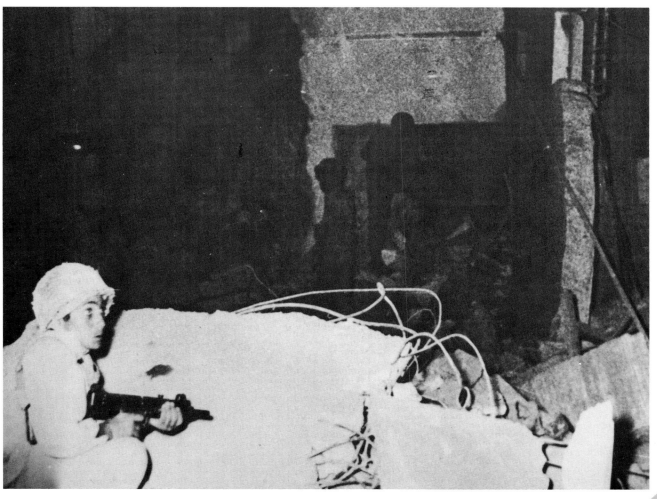

EARLY JUMPS

Israel's 1948 War was outstanding in military history as a confrontation where so much was achieved against so many by so few. Nevertheless, its many acts of extreme personal courage and initiative included very little commando action *per se*. A so-called commando battalion, led by Moshe Dayan, undertook some swift and successful actions against the Arab-held towns of Lydda and Ramla in Israel's Central Plain. This battalion, however, was more a mechanized/motorized infantry force than an actual commando unit.

On the other hand, the War of Independence marked the birth of Israel's Parachute Corps. Its founder, Major Yoel Palgi, was the only survivor of the ill-fated team that had parachuted into Eastern Europe in 1944 with Hanna Szenes. Now an officer of the newly created Israel Defense Forces, he was charged with the job of establishing a Parachute Corps for the IDF. The assignment, especially impressive for a mere major, had only one small hitch: nowhere in the State of Israel was there a usable parachute – and the Israel Air Force's inventory of aircraft suitable for dropping paratroopers was limited to exactly one.

Palgi, deciding to leave the problem of aircraft to the IAF, addressed himself to the question of parachutes. For the time being, making them in Israel was out of the question, and no foreign army was willing to sell usable parachutes to the IDF. Finally, an Israeli purchasing mission in Europe located a consignment of 4000 surplus British parachutes – about to be sold to a silk shirt factory – and immediately bought them, little realising the problems the defective chutes would cause their future users.

Once he had the parachutes and the aircraft – an ancient Curtiss C-46 Commando stationed at Ramat David, the same airfield where Palgi himself had trained – the new O/C Paratroops could establish his base. Having chosen Ramat David as the best site, there remained one problem for Palgi: where was he going to find troops? Once more the call for volunteers went up. This time, it was answered by an incredible variety of men not only from Israel, but from all over the world. Some were Holocaust survivors, determined to protect their new nation even though they knew nothing about armies; others were former members of the Palmach's 'Arab unit' – young Jews born in Arab countries who had served undercover during World War II. Several self-proclaimed French Foreign Legion veterans reported, boasting of three- and four-hundred jumps each... only to faint the first time they entered the Curtiss Commando! There was an American 'birdman', a Bulgarian parachute instructor, a genuine French paratroop officer; there were ex-6th Airborne paratroopers who had defected to remain in Israel. And many more.

Palgi now had over 100 volunteers who spoke at least a dozen languages; most of them had never seen a parachute. He had to mold them into a fighting unit – but first he had to teach them to jump. Using limited resources augmented by great ingenuity, the young major constructed a jump tower out of rope and an upended Bailey bridge. Another of Palgi's 'inventions' was a dummy aircraft, parked on Mount Carmel to give his recruits a feeling of height. However, the new O/C realized the limitations of his improvized educational aids. Something better, he reported to Ben Gurion, would have to be found. Ben Gurion took a look around – and came up with a human dynamo named Karl Kahane.

Kahane, born in 1900 – 18 years before Palgi – was a veteran in every sense of the word. He had fought in the two World Wars – in the army of his native Austria in World War I and with the British in World War II. Kahane's actions in the second Great War included distinguished service with a crack commando unit in Eritrea, fighting with Stirling's raiders in the Western Desert, and parachuting into Albania from 12,000 feet to help local partisans in the struggle against the Germans. To this considerable list, already especially impressive for a man in his forties, he now added one more honor: the post of Chief Instructor for the new IDF Parachute Corps.

Kahane and his assistant, a tiny but forceful Bulgarian named Sammy, enlisted the aid of the Czech Army in organizing a clandestine parachute training course in Czechoslovakia. Wearing Czech Army uniforms in camp and civilian clothes outside, some of the Israeli paratroopers-to-be were given a short but effective training course. However, despite the best efforts of all concerned, the course ended too late for the new paratroops to take part in the War of Independence. In addition, several serious accidents in the fledgling Corps had undermined the confidence of the IDF High Command. After the signing of the armistice agreements, the General Staff considered disbanding the unit completely.

But it was decided to give the paratroops one last chance. In 1949, following Yoel Palgi's resignation from the IDF, command of the Parachute Corps was awarded to Yehuda Harari. Harari, born in 1919, had fought with the Jewish Brigade and distinguished himself in Italy. Now he applied himself to the task of instilling discipline into the Corps. His first act as O/C was to review the personal files of his men, culling the worst offenders from the ranks. This done, he proceeded to establish an orderly training routine, moving the paratroop base to the Tel Nof airfield in the Coastal Plain and turning the Parachute Training School from an improvized facility to one worthy of the IDF.

Slowly, the paratroops became the first-rate unit Ben Gurion had intended it to be. In recognition of the fact, the 'Old Man' personally awarded the paratroopers' wings to the graduates of the first class under Harari – a 36-day course ending with a parachute jump over the beach at Jaffa, south of Tel Aviv. The wings, along with a maroon beret and red jump boots, soon became a symbol of prestige. Volunteers now flocked to the Parachute Corps; while some of them were mainly interested in the wings and beret, most joined for far more serious reasons: the honor of serving in an elite unit in defense of the State of Israel.

Once he had the men, Harari – by now a lieutenant-colonel – also got the aircraft. More Curtiss Commandos and some DC-3 Dakotas were acquired by the IAF, and a regular jumping schedule was worked out. The Parachute Corps was now a coordinated unit; its professionalism and polished appearance were drawing accolades at ceremonies and parades. GHQ, however, was still not satisfied. True the unit was impressive in the air and on parade. But would the Corps measure up in combat? The General Staff's uncertainty led them to turn down Harari's frequent requests that they use his troops in

some of the almost-nightly border actions of the early 1950s. Harari became more and more frustrated. His frustration filtered down to the troops under his command, and their performance suffered. This state of affairs would continue until 1953, when paratroop and commando warfare in Israel entered an entirely new dimension.

After the end of hostilities in 1949, many of the IDF's experienced commanders decided to resign from its ranks. The concept of professional soldiering in peacetime seemed alien to these men whose ideas of armed combat had developed during the years of the underground movements. Their resignation affected the quality of the IDF as a fighting force – and though the War of Independence was officially over, the young country was far from peace.

After the War and throughout the early fifties, Arab attacks on Israel's borders increased. Israeli reprisal actions were often unsuccessful, due to an unfortunate gap in the command stratum. Inevitably, GHQ began to lose confidence in its young army, which seemed to lack either the skill or the experience for the kind of quick, effective and well-coordinated reprisals necessary to retain control of the country's borders. It was therefore decided to create a special force of dedicated, motivated, first-rate fighting men – specially trained for just this type of action. The man selected to head the new unit was Ariel Sharon.

At the age of 25, Sharon had already become an accomplished military leader, having distinguished himself in the War of Independence as a battalion intelligence officer and company commander. In 1952, Sharon resigned from the IDF to study at the Hebrew University in Jerusalem. But the army was not about to lose such an expert officer. The following year, Sharon was asked by the O/C Jerusalem Brigade to volunteer, along with some of his comrades-in-arms from the 1948 War, to organize a reprisal action on the Arab village of Nebi Samwil, north of the capital. Gleaning his principles from the teachings of the eccentric British Captain Orde Wingate, leader of the SNS (Special Night Squads) of the thirties, Sharon instituted a rule that no one was to return

Early jumps, Israeli paratrooper style.

from this or any mission until it was completed according to plan. Under no circumstances were wounded or dead to be left on the battlefield. Following the Nebi Samwil raid, the brigade CO submitted a highly favorable report to GHQ. The General Staff read this with interest and decided that a man with initiative and military talent as great as Sharon's was just the type to command the prospective new unit.

Accordingly, Ariel Sharon was recalled to active service and given the job of creating an official counter-terror commando unit. He hand-picked his men from among his personal friends – most of whom, like Sharon himself, had left the IDF and returned to their peacetime occupations. He kept their numbers down – unit strength was a mere 45 – knowing that a small but truly elite force could do more in less time and at less cost than a large but mediocre unit. He trained them personally to the sharp edge of excellence, weaning them on strict combat discipline while preserving their outwardly unmilitary appearance. Within an astonishingly short time, the new outfit – given the designation of Unit 101 – was ready to go into action.

It took only a few weeks for the incredible combat successes of Sharon's commandos to become legendary throughout the IDF. Rapid-action raids in the Gaza Strip and across the Jordanian border consistently achieved their objectives, with minimum casualties to the Israelis. Volunteers from other units, as well as civilians, began to find their way into the ranks of Unit 101. One of these, a tall, delicate-looking youngster named Meir Har-Zion, was to become a legend in his own right.

Har-Zion, a loner and a trailblazer, had his first encounter with the Arabs at the age of 14, when his roaming through the Galilee led him into Syrian territory. Taken to Damascus for interrogation, he was subsequently released. Four years later, he joined the IDF; when he heard of the new unit, he requested a transfer to its ranks. For three years, from 1953 to 1956, Har-Zion set a shining example as a fighting man and a natural leader; in 1955, he was promoted to captain, although he had never attended an officers' course! His

skill and leadership helped him and the men of Unit 101 perform the impossible, time and time again, and escape unharmed. Finally, the gifted young officer was seriously wounded in action in September 1956; although his life was saved by an IDF surgeon operating under fire, physical disabilities forced him to resign. His name has to this day remained legend among Israel's commandos and paratroopers.

After several actions, most of them amazingly successful, Yehuda Harari succeeded in convincing GHQ to allow his paratroops to take part in the anti-terror campaign. Unit 101 was asked to carry out a reprisal action in cooperation with the paratroops – the destruction of an Arab village called Kibiya, just over the Jordanian border, from which murdering terrorists had struck the Jewish settlement of Yehud, killing a woman and her two children. Ariel Sharon was assigned command of the mission, which achieved its objective successfully – giving the General Staff, now headed by Sharon's close friend Moshe Dayan, the idea of incorporating Unit 101 into the Parachute Corps, which had by now achieved battalion-strength. Sharon was selected to command the overall force, and Yehuda Harari resigned from the IDF in protest – a great loss to the paratroops and to the Israel Defense Forces. But the Parachute Corps, under Sharon, was to become an elite fighting unit for the first time.

Early jump training.

ACTION AT MITLA PASS

1659 hours, 29 October, 1956. Twelve Israeli Dakotas and twin-engined Nord Atlas transport aircraft skimmed low over the hills of the Sinai Desert, dodging Egyptian radar surveillance. Approaching the Parker Memorial at the east end of the Mitla Pass, the formation climbed stiffly to 1500 feet. Once at altitude, Rafael (Raful) Eitan, battalion commander in what was now the IDF 202nd Airborne Brigade, was the first to jump, followed by 393 more paratroopers. The Sinai Campaign was on.

Eitan, a former infantryman (and future IDF Chief of Staff), had been transferred to the paratroops the previous year as a company commander. His excellent record ensured his rapid promotion. Now he was dropping out of the sky, 300 kilometers behind enemy lines – to be welcomed with cheers and black coffee by a party of Egyptian roadworkers who mistook his force for the Egyptian Army on maneuvers. Not bothering to disabuse them of the idea, Eitan organized his men and settled down anxiously to wait.

Ariel Sharon was also anxious. His 202nd brigade, minus Eitan's paratroopers, had to link up with Raful's battalion in the middle of the desert. The overland route from the border to the Mitla wound through no less than three heavily defended Egyptian strongholds. And, to make matters worse, Sharon had no way of knowing when or how many enemy troops would reach the area now held by Raful's men. Sharon's mission was one that would have daunted many lesser men, but the brigade commander was confident of his unit.

As Eitan's battalion hit the rocky sand, the rest of Sharon's brigade rolled out in a long column of half-tracks, trucks and jeeps, supported by a company of AMX13 light tanks. Without incident, they took the first Egyptian outpost, Kuntilla – scene of a reprisal raid some months before; the Egyptians fled as the paratroopers approached.

The next post, Themed, was not so easy. It was dawn by the time the IDF forces reached the objective, which was surrounded by a dense minefield and a considerable defense perimeter. Aharon Davidi's battalion was given the job of clearing the position. Attacking from the east with the sun behind him, Davidi

drove his halftracks straight into the compound. In forty minutes it was all over, with fifty Egyptian dead against four IDF casualties. The third Egyptian position, Nakhl, was stormed that afternoon.

Almost exactly 24 hours after the onset of the campaign, the two forces linked up and reconnoitered in the sands east of the Mitla Pass, Sharon rapidly appraising the situation and informing Eitan of the past day's developments. Elsewhere in Sinai, the 7th Armored Brigade – Sharon's archrival within the IDF – was busy winning its laurels in highly successful combat. But here, with no enemy troops in sight, the 202nd Airborne appeared to be out of a job.

Contacting GHQ, the restless brigade commander asked to be allowed to cross the Mitla Pass and push westward to the Suez Canal. Infuriatingly, permission was denied. Sharon, never one to sit idle when action was possible, then asked permission to mount a reconnaissance patrol into the pass. This was granted, provided the patrol withdraw at the first encounter with enemy resistance. Sharon reluctantly agreed; he then had to break the news to his battalion commanders, by

The paratroopers prepare the site for defense.

A paratroop sergeant at the Mitla armed with an Uzi SMG and equipped with an American steel helmet and IDF standard combat webbing.

The only contact with friendly forces in the rear was through the communications set. Constant air cover was maintained by IAF Mystere IV fighters which shot down an Egyptian MIG, whose smoke is seen in the background.

now as impatient as their leader.

Most anxious was Motta Gur, commanding a battalion of NAHAL paratroops. (NAHAL, a Hebrew acronym for Fighting Pioneer Youth, is composed of young men and women who spend a portion of their military service establishing and maintaining agricultural settlements on Israel's borders. NAHAL men, however, spend most of their service as volunteers in one of the IDF's elite units – including the paratroops.) Gur's battalion had been chosen by Sharon to lead the way into the pass. First into the Mitla was a young lieutenant in a halftrack; Gur himself occupied the second of the battalions' six halftracks, followed by 120mm heavy mortars, supply trucks carrying the recce company, and three AMX tanks.

The battalion had been briefed not to expect too much resistance; scanty intelligence available at Brigade HQ, supplemented by reports from the IAF, which had attacked in the area the day before, indicated no strong Egyptian troop concentrations near the pass. Nevertheless, Sharon did not discount the possibility of an armor attack, either from Bir Gafgafa to the north, or by retreating Egyptian tanks pushed westward by the 7th Armored.

Suddenly, as it rounded a bend, the young lieutenant's halftrack was fired on by numerous Egyptian weapons. The driver and commander were hit by the first volley; out of control, the halftrack turned sideways and blocked the road. Gur, in an effort to save the wounded

men, led the rest of the column into the fray; his own halftrack was hit by anti-tank fire inside the pass. The battalion CO, rallying the survivors, took cover in a shielded spot beside the road. Meanwhile, Sharon's second-in-command, accompanying the lead force, ordered the halftracks following Gur forward through the pass. This left the battalion hopelessly divided, with no possibility of contact between the halftracks and tanks that had thundered through the pass, Gur's force – a few men, one tank and some LMGs – inside the pass itself, three damaged halftracks and an ambulance full of wounded not far from Gur, and the rest of the battalion still at the eastern opening.

Rather than the few isolated Egyptian troops Intelligence had led them to believe awaited them, the enemy consisted of an entire infantry battalion, the 5th, of the 2nd Infantry Brigade. Reinforcing the battalion were 14 heavy machine guns, 12 6Pdr AT guns, 40 Czech-made self-propelled guns, and many Alpha light machine guns. Taking advantage of the topography, all of these had dug into the hillsides of the narrow defile. From concealed positions in caves and crannies, they rained fire on the stricken paratroopers. To make matters worse, four Egyptian Air Force Meteors now attacked the columns jammed at the eastern opening of the pass, knocking out the heavy mortars and supply trucks, and causing considerable casualties.

The indomitable Davidi now saved the situation once again. Ordering the recce company onto a hill north of the Mitla, he directed them to fight their way downhill and reach Gur and his men, who were still stranded inside the pass. The company, spread out into platoons, began to clean out those enemy positions that could be seen. Suddenly, fierce fire from the south side of the pass tore into the company; the men, incapable of locating the enemy positions, were unable to fight back. Meanwhile, Gur was importuning Davidi by radio to speed up the company's advance.

As the recce company approached the edge of the hill, they came upon an almost vertical drop. Peering down, the company commander realized that the Egyptian positions were actually cut into the side of the mountain! Informed of the fact,

Perimeter set around the DZ.

After the drop, the paratroopers assemble with their gear.

Tank recce patrol checks their position near the Dayka pass, south of Abu Ageila. This unit located the pass which allowed the 7th Armored Brigade a tremendous victory in the Sinai Campaign.

Davidi decided to ask for a volunteer to drive into the pass and draw fire, enabling the paratroops to pinpoint the enemy positions. The call was answered by many brave men – including Rafael Eitan and other senior officers – but Davidi chose a young driver to undertake the desperate mission. The courageous soldier selected an unarmored jeep and drove at full speed into the pass, to be riddled by countless bullets within seconds. His sacrifice enabled the paratroopers to pinpoint the Egyptian positions and prepare to take them out.

As darkness fell, small Israeli task forces inched along the steep ridges, lobbing hand grenades into the caves sheltering the Egyptian weapons. By midnight, the battle was over. The 202nd Airborne regrouped; counting their forces, they were shocked to find they had lost 38 men, with a further 120 wounded – many of them irreplaceable commanders who had forged ahead of their units into the fiercest fighting. But not a man remained of the Egyptians: more than 260 enemy soldiers had been killed.

The battle for the Mitla Pass was the first major combat action of the 202nd Airborne Brigade. Faced with an ambush of such ferocity and strength, a lesser unit would have crumbled. But the training, initiative and extreme courage of the IDF

Parachute Corps had saved the day. The Mitla action went down in history as the first of many tests which Israel's paratroopers would pass with all honors.

The paratroopers disassemble a paradropped jeep as they organize at the Mitla DZ.

The paratroopers dig in at the DZ.

NIGHT RAIDS

Raid on Nukeib, 1962. Another wounded IDF soldier being carried back to the evacuation point.

Nukeib – battalion commander Zvi Ofer gives first aid under Syrian fire to one of his soldiers.

Mopping up Syrian opposition in the trenches at the stronghold overlooking the Sea of Galilee.

Raid on Samua, 1966. The target is blown up during the raid.

PARATROOPERS AT WAR – 1967

The years between 1956 and 1967 were relatively peaceful ones for the Parachute Corps. These were years of reorganization, of strengthening, of ripening. Combat techniques were analyzed and perfected; more battalions were created, raising the rank and responsibility of many junior commanders. However, action itself was scarce. Several minor clashes on Israel's northern and eastern borders involved paratroops; only one raid, a sortie into the Jordanian village of Samua shortly before the 1967 war, could be called a major action.

Three paratroop brigades faced the Six Day War. One, a reserve unit commanded by Colonel Danny, a veteran of the Gaza action 12 years before and famous for his flowing beard, was assigned to the division led by the former paratroop o/c, Ariel Sharon. Colonel Danny's task was to attack and destroy artillery positions deep inside the Abu Ageila complex, a Soviet-style linear defense position occupied by the Egyptian 2nd Infantry Division. In a surprise attack, Danny's men were to be heliported to within four kilometers of the Egyptian batteries – the first brigade-scale airborne operation in IDF history. From there, they were to pounce upon the Egyptian artillery, preventing them from

Brigade command vehicle breaking through the Lion's Gate.

Evacuating the wounded.

Mopping up near the Damascus gate.

firing on the rest of Sharon's division, which had been forced by ground and defense positions to attack the 2nd Infantry complex frontally.

As the last light faded, the lead helicopter dropped the advance party over the landing zone. The paratroopers quickly marked the area for the choppers yet to come; they did not have long to wait. Within minutes after the area was marked, the first big bird hovered over their heads, releasing the lead contingent of troops, who disappeared into the darkness. As one helicopter left the area, another appeared, discharged its load of fighting men, and sped off in turn. Amaz-

ingly, the landing was undetected; by the time the brigade had assembled, the rest of the division had already begun to attack the Abu Ageila complex. Colonel Danny rapidly dispatched his men toward the artillery positions, over a nightmare obstacle course of sand dunes as high as 15 meters.

Nearing the artillery across the Abu Ageila road, Danny and his officers directed the troops to their objectives. Each Egyptian gun was assigned to a platoon of attacking paratroopers. The command was given, the troops charged in – and all hell broke loose in an incredible pyrotechnic display of exploding ammunition dumps and guns, scattering the Egyptian forces to the four winds. A call went out for enemy reinforcements; these, as they arrived, were ambushed along the road by IDF blocking parties, and torn apart by

Paras at Abu Ageila. In a coordinated move, these men took the Egyptian artillery positions in a surprise heliborne attack at night, the first such operation carried out by the IDF.

73

mines, grenades, and mortars long before they reached their objective. Within an hour, all the Egyptian batteries had been destroyed or abandoned. No more shells fell on Sharon's division.

On the day before this night battle, a regular paratroop brigade commanded by Rafael Eitan – the former battalion commander first to jump into the Mitla Pass in the Sinai Campaign – fought an impressive battle, as mechanized infantry, with General Tal's armored division at Rafah Junction. This heroic action – which itself could fill a book – will be dealt with in detail in another volume of our series, dealing with the Six Day War. Meanwhile, the third paratroop brigade, a reserve unit under the notable Motta Gur, had been left jobless. Originally assigned to parachute into El Arish and capture the crossroads, and then to link up with Tal's division along the coastal road, Gur's

The lead unit fighting their way through the narrow streets.

Motta Gur, CO 55th Paratroop Reserve Brigade.

Para AT support unit with M106 jeep-mounted recoilless gun crossing the lines near the Police Academy.

brigade was abruptly informed that the armor had captured the crossroads and the town of El Arish unaided. To the brigade's consternation, the mission was canceled; but it did not have long to wait for a new assignment. The Jordanian Army had entered the war despite an appeal by the Israel Government to refrain from hostilities; it was bombarding the Israeli half of Jerusalem with heavy artillery fire. A decisive action was needed – and Motta Gur's brigade was chosen for the capture of Jerusalem's Old City.

Paratroopers fighting their way into the Old City, taking cover near the Lion's Gate.

Paratroopers fighting in Jerusalem.

JERUSALEM

Switching missions without extensive preparation is at best a hazardous job under any circumstances. To go into battle totally unprepared and without a complete set of orders required a truly first-rate unit – such as Gur's paratroop brigade, composed mainly of Mitla Pass veterans. The problems Gur faced were formidable. No Israeli had seen the Old City at close hand for 19 years; the approach to the objective provided only scattered glimpses of walls and towers. The concentration zone was under constant bombardment; there were many casualties before the engagement actually began. Nevertheless, Gur did not lose his nerve; as darkness fell, the reserve paratroop brigade awaited the order to move out.

The lead battalion, the 66th, led by Lt. Col. Yossi Yaffe, was ordered to occupy the Jordanian Police School, a formidable bastion only 200 meters from the border. Yaffe would then take the Ammunition Hill position preparatory to linking up with the 71st Battalion; after capturing the Ambassador Hotel at Sheikh Jarrah, north of the Old City, he was to proceed to Mount Scopus. Gur himself placed the brigade HQ on the roof of a building not far from the breaching area. Signalers set up telephone and radio equipment with antennas at low angles to prevent detection. The clock hands inched forward – midnight, 1 AM, 2 AM. The shells shrieked overhead.

0215. The signal for attack is given. The silence is taut, unbearable. Suddenly a streak of bright red fire cuts through the darkness, followed by a dull thunder. The first breach has been made. Now charges explode one after another. The enemy fire intensifies as artillery concentrations are zeroed in on the Israeli troop assembly zones. Now Gur's artillery commander begins to implement *his* fire plan, shifting his batteries from target to target. 155mm and 160mm shells rain down on the Police School; two batteries of 25-pounders plough into the Jordanian trenches on Ammunition Hill. The whitewashed walls of the police fortress glow hauntingly in the light of two giant searchlights. The supporting ranks rumble into action, firing point-blank at their targets. Machine gunners smash down the doors of evacuated houses along the borders.

Then, taking up new positions inside the buildings, they fire out the windows at the Jordanians. Mortar fire resounds in the congested area. Casualties begin to mount.

0230. Brigade artillery fires a retaliation barrage. Commanders leap forward toward the first border fence, their troops hot at their heels. The lead commanders, squatting, direct the positioning of bangalore torpedoes; each man springs forward, hooks up his section of the tube and pushes. The assault squads begin to take cover as the platoon leaders light the charges and rush back under heavy mortar fire. Two seconds, three – then a series of tremendous explosions rips the fence to shreds. Each platoon leader takes his position at the right side of the breach his men have made; platoon sergeants squat on the left, illuminating them with small, unearthly-looking blinker lamps. Rows of crouching figures slip through the breach and approach the second fence; it, too, is rapidly broken through. The enemy has now located the breach and is spraying the area with withering machine-gun and mortar fire; medics run to move the wounded to safety. A third fence is breached, then a fourth, and the

battalion pours into the Police School. Hand grenades, bazookas, and recoilless rifle shells explode around. Guides with blinker lamps placed at vantage points direct the men to their objectives.

B Company then left the Police School for the rest of the battalion to mop up, and pushed forward to its next objective, Ammunition Hill, behind the police compound. The lead squad charged through the connecting trench, rapidly arriving at the road. A burning truck illuminated

Facing page:

Paratroopers fighting the Jordanians at Ammunition Hill.

IDF paratroopers seen in the maze of trenches at Ammunition Hill.

Grenade! – the paratroopers mopping up Jordanian opposition on the outer perimeter of Ammunition Hill.

buildings on both sides with an eerie orange glow. The paratroopers rushed into the first building, spraying the corridors with submachine-gun fire and tossing hand grenades. Quickly clearing the rooms, they broke through to the far side of the building. Each platoon took position in a Jordanian trench and continued fighting. But the trenches were narrower than those in which they had been trained; terrible congestion ensued. More and more men leaped into the dark

trenches, unaware of the pile-up at the far end. Meanwhile, Jordanian mortar fire pounded the Israelis.

A platoon commander took the initiative; ordering the men already in the trenches to lie low, he pushed the others forward over their comrades. Movement was soon restored and the attack continued at full force. The Jordanians, however, rallied unexpectedly, turning the area into the fiercest battleground the paratroops had ever known. The IDF troops charged along the trenches, firing their Uzis in short bursts and slowly gaining ground. But the cost was high, with terrible casualties. Dawn found the exhausted survivors deep inside the enemy stronghold, still under heavy fire.

At this point the battalion second-in-command, realizing the intensity of the fighting, pushed some Sherman tanks into the compound. The paratroopers, encouraged by the sight of the Shermans, directed them into position. Highly accurate fire roared from the tanks, decimating the Jordanians. But the fight was far from over. Hardly visible, an especially deep Jordanian bunker loomed ahead. The Arabs inside defended it tenaciously, mounting counterattacks and ambushes against the men of B Company as they rushed into the area. It was then that a junior officer realized the only way to destroy this bunker was to blow it up. Tossing demolition charges to a man near the wall, he hurriedly instructed the soldier to place them in position. The officer then rushed in, lit the fuse, and leaped to safety. Seconds later, a tremendous explosion shook the whole of Ammunition Hill. Despite 21 Israeli dead and more than half the company injured, the position was finally secure. All but three of the Jordanians – more than 100 men – had been killed.

Realizing the large number of objectives to be covered by Gur's brigade, Yossi Yaffe intentionally did not inform the brigade commander of the seriousness of the situation on Ammunition Hill. Gur therefore did not dispatch reinforcements to the area, but kept his other battalions attacking their original objectives – all of which had been reached by morning. Gur now asked and received permission to press on into the Old City itself. Attacking through the Lions Gate, Gur's brigade rapidly wiped out all remaining Arab resistance to achieve the 19-year-long dream of the Israeli people: the liberation of the Western Wall and the Temple Mount.

GOLANI – LIGHTNING STRIKES AT TEL FAHAR

On the afternoon of 9 June, 1967, Barak (Lightning) Battalion of the Golani infantry set out to capture Tel Fahar, a heavily fortified Golan Heights stronghold making up part of the Syrian forward defense compound and guarding access to the northern Golan slopes. It was manned by a company of the 187th Infantry Battalion, supported by several 57mm M43 anti-tank guns, two B-10 recoilless guns, three twin-barrel Guryonov machine guns, and a dug-in battery of 82mm mortars – a deadly combination indeed.

The hilltop position, hewn into the rocky mountainside, overlooked Tel Azaziat, one of the strongest fortifications some 1500 meters to the northwest, and a smaller outpost, Burg-Bavill, straddling the west-east track and guarding the southern minefield. Farther up the mountain slope to the east, in the fortified villages of Ein Fitt and Zaoura, were situated the main elements of the Syrian 11th Brigade.

Tel Fahar, the strongest and best defended Syrian position in its northern defense complex, was undoubtedly a very hard nut to crack. It was situated on two hillocks, northern and southern, dividing the position and supporting one another with extensive trenchworks running throughout the fortification, surrounded by several wire and mine obstacles; top-covered bunkers overlooked access routes, which all ran north-south and were bottlenecked by minefields.

Golani was to follow the 8th Armored Brigade up the slopes along a specially prepared track; while the armor shifted to the southeast, Golani's M3 halftracks and the M50 Sherman tanks supporting them were to turn north and northeast to storm the Syrian 11th Brigade complex and open the northern route running along the Hermon slopes towards Masadeh.

Pushing its two forward battalions up the slope, Golani reached the turnoff north of Na'amush. The spearhead battalion raced along the Syrian patrol track and stormed into Tel Azaziat from the rear, capturing it after a sharp battle. Barak Battalion was briefed to assault Tel Fahar from the east using the TAP-line (Trans-Arabian oil Pipeline) road but, missing the hard-to-identify connecting track, the unit turned northeast

onto an improvised dirt road and came under heavy fire from Syrian positions. Trouble hit the battalion immediately and trouble was to stay with it until evening.

Hit in quick succession by anti-tank fire, three Sherman tanks now blocked the road. Desperate to keep moving, the battalion commander, Lieutenant Colonel Moshe Klein, crashed over the boulder strewn slopes and, with several vehicles following, regained the track. Now heavy artillery concentrations exploded among the hapless vehicles filled to the brim with Golani fighters. One M3 was hit and its passengers, some severely hurt, scrambled for cover among the rocks. Communications were disrupted as the survivors raced forward, and many individual dramas developed in their wake.

As the vanguard reached the base of the hill, it became clear that the assault on Tel Fahar was being undertaken at its strongest point instead of its rear, as had been originally intended. Now, fully engaged and under fierce fire from three sides, there was little choice but to attack, despite the fact that the battalion commander had lost communications with his forces. Capturing Tel Fahar was to be at the initiative of junior officers. As it turned out, the action was to be one of Golani's finest, clearly demonstrating the high quality of its fighters and junior commanders who won the day against a stubborn and courageous enemy who gave no quarter.

While their vehicles were exploding

S-58 carrying troops to the Golan.

In the Golan, just after overcoming a Syrian position overlooking the Huleh Valley.

along the dusty track, the survivors assembled and the wounded were tended by medics working feverishly under deadly fire. 'A' Company reached Tel Fahar's outer defenses with 25 men out of its original 60. Assaulting in two sections, each led by an officer, the men raced panting up the slope pursued by flat trajectory fire from the Syrian Guryonov machine guns. Throwing themselves on the outer fences, one man fired a rifle grenade into the bunker slit visible just over the top,

Mopping up on the lower Golan slopes.

and silenced it. Soon the Golani fighters were inside the trenches, racing forward and firing from the hip at anything that moved.

Splitting into even smaller forces, with junior officers leading, the attackers stormed through the dark fortifications. Losses started mounting as the Syrians savagely fought back, their bunkers spitting fire while the ground shook under artillery and mortar concentrations.

By now the battle had broken down to the individual level, with groups of two and three fighting for their very lives. Led by a young corporal, four men stormed

the main bunker – no orders had been given, nor were they needed. The bunker spat lethal fire and contact was imminent. Racing forward, two men were hit and fell. The other two, firing all the way, reached the bunker and hurled two grenades into its slit, storming into the darkness immediately after the explosions.

As they emerged and raced along the connecting trench, they suddenly came face to face with their battalion commander, Lt. Col. Klein, who was alone. Assembling a few men around him, Klein ordered them to follow him. He ducked and ran forward, shooting a Syrian who suddenly cropped up in his way. Klein then raced with his men to the northern sector of the position, where, hit by a sniper, he fell, mortally wounded. The men around him gasped in terror. Then, out of nowhere, an artillery captain, who had followed the CO up the hill, appeared and rallied the terrified men, racing them into the trenches and forward. By now more Golani fighters had joined in the fight and, mopping up the Syrian survivors, secured the southern compound.

There was still heavy fighting in the northern sector. Urgent help was needed, as small sections fought individual combats, with junior officers trying frantically to coordinate actions in the heat of battle.

One officer raced downhill and established contact with brigade HQ using a tank radio. Soon, the Sayeret Golani recce unit raced from the southeast into the northern compound. With not a moment to lose, the Commander, Captain Ruvke, stormed into the trenches with his men closely in his wake. Ruvke grappled with a Syrian officer who had brandished a pistol. A Sayeret member killed the Syrian and the group stormed on. By now the fight for the northern compound was reaching its climax, with the Syrians firing for all they were worth. Trying his best to coordinate the battle and signalling to the soldiers in the southern compound, Ruvke and his men moved on. After savage hand-to-hand fighting, they finally succeeded in capturing the northern hill.

The battle for Tel Fahar had taken until evening when, in a final act of triumph, the weary Golani fighters raised their flag in the torn branches of a tree. Twenty-two men, including their commander and many of their officers, had died on that bloody hill, but their mission was completed and the road east was open.

DEEP PENETRATION RAIDS

Israel's success in the Six Day War stretched the country's borders so that they were much longer than they had been prior to 1967. This, in turn, meant that they were much more vulnerable to guerrilla infiltrators – especially in the Jordan Valley. The terrain on both sides of the Jordan made it easy for infiltrators to slip across, hide out in relative safety, and then smash through to targets inside Israel. The Parachute Corps was assigned the task of securing Israel's borders and preventing guerrilla forces and terrorists from reaching Israel.

This was a far different task from those assigned the Corps in the Sinai Campaign and the Six Day War. There was no longer a question of organised, pitched battles against a large-scale regular army. Instead, the paratroops had to play a game of patience, a war of will and wits. The enemy now took the form of countless small groups, crafty and well-informed – many of them born inside Israel's borders – who cared less for their own lives than for the 'cause' of exterminating as many

Jews as possible. The years from 1968 through 1970 turned the Jordan Valley into the scene of a never-ending manhunt. In the long days and nights of those three years, many commanders were killed leading their troops in this new and deadly game. But after three years of heavy Israeli casualties to the Palestine Liberation Organisation, the terrorists seemed to lose heart.

IDF patrol on the banks of the Jordan River. The dense vegetation has provided shelter for infiltrators since the days of the Bible.

When tracks are found, the search-and-destroy operation commences; chasing terrorists across the Jordan River.

Meanwhile, an entirely different kind of attrition was going on in Sinai, on the banks of the Suez Canal. Daily artillery barrages rained hundreds of tons of red-hot steel on the Israeli defense positions, taking an ever-increasing toll among the soldiers stationed there. The situation rapidly became unbearable, and the IDF decided on action. The only solution that could effectively convince the Egyptians to stop firing involved a series of audacious raids deep into enemy territory. These intrepid missions, some of which gained world fame, others still top secret, were awarded to the Parachute Corps.

Among the first of these raids was a lightning action against the Naga Hamady transformer station, some 250 kilometers north of the Aswan Dam. The station and a series of bridges over the Nile further north were blown up in tremendous explosions which apparently shocked the Egyptians into silence for several months. This blow was struck by two paratroop raiding parties, ferried by heavy helicopters over the Gulf of Suez under cover of darkness to simultaneously hit their targets. All raiders returned unharmed.

But the fighting along the canal was not over. After a few months of relative calm, the shelling started again, even more fiercely than before. GHQ finally, reluctantly, decided to launch the IAF against the harassing enemy. Before this could be accomplished, however, a fortified anti-aircraft position at the entrance to the Suez Canal had to be knocked out. This bastion, named Green Island, had been constructed by the British during World War II on a small coral reef; it now included surveillance radar, AA guns, and an armed garrison to defend the position.

The 145 meter long, 50 meter wide ex-British fortress on Green Island, four km south of Suez. The fortress mounted 85mm anti-aircraft and machine guns and a radar installation. It was manned by a company of Egyptian infantry. A surprise raid by IDF Naval Commandos succeeded in blowing up the fort.

Carrying the chase into Jordan; in front of a cave to where the tracks led. The terrorists are soon overcome with smoke and explosives.

The search for infiltrating terrorists; Jordan Valley. Sappers check a banana tree for mines planted by terrorists against the villagers.

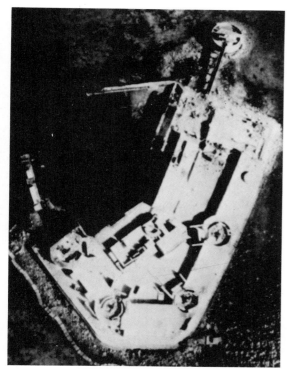

On a pitch-black night, a raiding party of paratroops and marine commandos led by Zeev Almog, future o/c Israel Navy, was brought in from the sea. As the leader climbed the slippery wall to the connecting bridge, he came up against an Egyptian sentry. Rapidly disposing of the guard, the attack party swung over the rail and crossed the bridge to the main position. Running swiftly along the wall around the installation, the raiders tossed grenades into several chattering pillboxes and vaulted down into the interior court. They proceeded to clear the buildings prior to laying demolition charges. But meanwhile, Egyptian shore-based artillery, aware of the raid, began to blast the island. Hundreds of shells rained down on the paratroopers, who retreated into the

night carrying their six casualties. As they left the facility behind, the hastily-laid demolition charges blew, lighting up the Gulf of Suez. Mission accomplished.

The paratroopers' reprisals against terrorism sometimes carried them far beyond the borders of Israel. In the fall of 1968, the PLO launched several attacks against civilian aircraft, including a notoriously deadly assault on an El Al airliner at Athens Airport. After lengthy deliberation, the Israel Government decided to launch its retribution against Beirut Airport, take-off point for many of the skyjackers.

On the evening of 28 December, 1968, Beirut Airport was humming with activity. The tarmac was studded with airliners, the terminal crowded with passengers. Sud-

Shadwan Island as seen from the air. The target is marked by the lighthouse around which were located the Egyptian positions.

December, 1969, an even more audacious paratroop raid resulted in the capture of a brand-new, Russian-made P-12 radar installation, recently set up by Soviet crews to guard the west bank of the Gulf of Suez.

denly, four Super Frelon helicopters appeared in the sky above the terminal, then landed to disgorge a team of paratroopers. Not a shot was fired as the IDF raiders took control of the terminal, urging all civilians inside. A fifth helicopter, hovering over the Beirut highway, stopped all incoming traffic; the outgoing lanes, left clear, immediately filled up with escaping Lebanese cars in a bumper-to-bumper traffic jam. The demolition crews swarmed over the tarmac, picking out those aircraft owned by Arab airlines. Rapidly, they placed explosive charges under the landing gear of the aircraft – fourteen of them – rendering them useless in the explosions that followed. Neither the attackers nor the civilian onlookers suffered any casualties.

Almost exactly a year later, on 26

Forward IDF paratroop HQ observing the target.

RAS GHARIB

Towards the end of December 1969, the War of Attrition between Egypt and Israel was reaching its climax. The Israel Air Force continuously attacked Egyptian targets along the Suez Canal. The Egyptian air defense, badly damaged by the attacks, received new impetus from the Soviets, who flew in reinforcements and new, advanced equipment.

Among the targets attacked by the IAF fighter-bombers was a fortified radar site on the Gulf of Suez near Ras Gharib, containing two radars, one of them a long-range static station which was knocked out. The second radar, a mobile unit well dug-in, remained intact, as recce photos taken after the raid showed. Heavy anti-aircraft defenses were located in the vicinity of the site, including artillery and heavy machine guns. Some three months earlier, in September 1969, an Israeli armored task force, landed on the Egyptian shore by amphibious craft, had destroyed a large Soviet-made BARLOCK GCI radar station at Ras Zaafrana, 70 kilometers to the south. IAF radar experts, examining the Ras Gharib aerial photo, identified the surviving radar as a new Soviet P-12 SPOONREST A. An earlier version (KNIFEREST P-10) had been captured, badly damaged, in the Six Day War. The new P-12 was known to be an extensively modified version, of which details were completely unknown to Western experts. From the information available it became certain that the P-12 radar was designed to detect low-flying aircraft – an activity at which the IAF pilots excelled in attempting to overcome the growing efficiency of the Soviet-bolstered Egyptian air defense then building up along the Canal. It was therefore vital to either eliminate the flanking radar coverage along the Gulf of Suez or at least obtain detailed information on the technical characteristics of the new system.

The IAF was already implementing a large-scale air defense suppression plan, and the immediate reaction to the location of the P-12 site was to destroy it with an air strike.

As he interpreted a new recce photo, a young IAF intelligence corporal had his own idea. He asked his superior why the site had to be destroyed instead of bringing the radar intact back to Israeli territory on the other side of the Gulf, enabling

its contents to be closely examined. The officer had a word with the IAF Chief of Operations, who immediately started things rolling. The air strike already laid on to destroy the radar site 23 December at 1000 was cancelled.

A new idea quickly took shape during a meeting in which senior IAF and paratroop commanders discussed the technical and tactical aspects before submitting a plan for taking out the P-12 radar to the Chief of Staff, Lieutenant General Haim Bar-Lev. A planning group was ordered to work out details; it was led by future Chief of Staff Colonel Rafael Eitan, who would command the airborne forces.

It was decided not to lose any time in implementing the mission, and the night of 26 December was chosen, 'H' hour being 2100.

As the paratroopers started training on a model site, the air force team entered into the intricacies of transporting such heavy equipment by helicopter. The P-12 SPOONREST was known to consist of two shelters mounted on standard ZIL-15 trucks, of which many hundreds had been captured by the Israelis in the Six Day War and were now rotting in salvage depots. In order to test the best way to lift the shelters, air force officers ordered two ZIL truck and shelter combinations to be towed to an IAF airbase for trials.

Searching for information among the available sources, IAF intelligence found that the SPOONREST station included a main operational shelter, in which the radar set, control and display consoles were mounted. It was estimated that this shelter would weigh about 3.5 tons if dismantled from the truck. The second shelter, including the antenna assembly and

its electrical motor unit, would weigh about a ton less. It was therefore vital to dismantle the shelters from the trucks in order to lift them off, and a field trial was immediately scheduled. Experienced helicopter pilots were called in to calculate the weight lifting capacities of available choppers.

The Sikorsky CH-53D, recently arrived in Israel, was ideally suited for the mission. With a maximum weight of 19 tons, its lifting capacity under normal operational requirements would enable it to do the job. However, the mission was far from normal and special allowances as to fuel capacity, temperature environ-

Sikorsky CH-53D helicopter.

The Yagi antennae of the P-12 SPOONREST.

ments, and the positioning of the P-12 site had to be taken into consideration. Furthermore, the weight of the dismantled shelter was far from precise; at best, it was an educated guess. Having considered the matter thoroughly, the CH-53 helicopter commander decided to put his trust in the Sikorsky designers, carefully weighing his chances on the safety margin which is usually included in the producer's manuals. It was a calculated risk – a touch-and-go decision – but someone had to take it.

By then, the paratroopers, selecting a team of qualified welders from their ranks, had started to dismantle the shelters from their trucks. Once clear, the empty shelter was filled with rocks to weigh exactly 3.5 tons and, with special rigging fitted by the airborne transport section, the shelter was made ready for its first heliborne lift; as the CH-53 chopper hovered overhead, the load was hooked up. Now came a tense moment as the pilot carefully hoisted the shelter – it worked without a hitch, to the relief of the senior

The control console on its arrival in Israel.

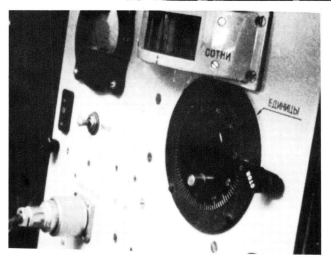

Detail of the installation.

officers watching below. The trials were repeated several times, the large chopper carrying the weight around the base with obvious ease.

Now the paratrooper team went ahead practicing their dismantling technique, aiming to cut down to an absolute minimum the time required. After several hours, the shelters were being dismantled in less than half an hour's working time, an impressive feat anywhere.

Meanwhile, assault team commander Lieutenant Colonel Arie Zimmel, leading the crack NAHAL paratroop unit, was bent over maps and aerial photos with his staff, working out the final details of the plan.

The Egyptian radar site was located not far from the shore, opposite the Israeli-held oilfields of Abu-Rodeis. The ZIL trucks and shelters had been placed in dug-outs surrounded by protective earthworks. Defending the site was a section of infantry, as well as the radar operators, with all-round defenses and several tents serving as quarters. A few kilometers

away, the Egyptians had substantial forces including a mobile reserve unit located further south along the coastal road. Due to recent Israeli activity along the Gulf, especially since the armored raid, the Egyptians were on full alert status, more so after nightfall, when they expected Israeli action.

To complete the assault party, a radar expert who had examined the P-10 KNIFEREST radar and was familiar with captured Russian electronic equipment was asked to volunteer for the mission. Having no combat experience, the airman joined the paratroop command group in a rehearsal exercise. Heavily laden with unfamiliar gear, the man, panting and sweating, stuck to Colonel Zimmel's party as they force marched over the hills near their base. But he made it, earning a good word from the eagle-eyed commander who watched his progress closely.

As 'H' hour approached, the assault party was flown in· Sud Aviation Superfrelons to a forward base at Ras Sudar on the Gulf coast, where the men rested. The airborne brigade commander, Colonel Haim Nadel, gave his final briefing and then set off to join the Chief of Staff who, with several senior army and air force officers, had established a forward command post near Abu Rodeis. By nightfall, the assault team had made ready and boarded their helicopters. The CH-53s, nearby on an airstrip, waited for the word, the pilots tense.

In the dark, the Superfrelons, completely blacked out and with control panel lights carefully dimmed, took off and flew low in wide formation, skimming the waves of the Gulf and making straight for the Egyptian coast which they reached safely after a short flight. As they crossed the coastline, IAF fighter bombers started a diversion raid on the Egyptian military camps south and north of the Ras Gharib radar site, making lots of noise, while IDF long-range artillery started bombarding other targets to keep the Egyptians busy.

Now the helicopters arrived over their designated landing zone and groped for a visible reference point on which to land. In the dark this was far from easy, even for an experienced chopper pilot. As they hovered, the pilots clearly saw the dimly lit radar site some seven kilometers to their north and many explosions from the bombing raid to their south. There was not a moment to lose. So far, it seemed their presence had not been detected.

Groping for a landing, the helicopter leader tried to set down his heavy chopper, with sand and dust flying up as he settled. On a snap decision, the pilot used the Egyptian radar antennae, barely visible, as a reference point; it helped and he got down. The other two less experienced pilots had trouble, and made several abortive attempts to land. Sending a paratroop officer to signal them in, the chopper leader got them all down safely and the assault team quickly assembled and set out in silence towards their respective objectives. Navigating in the darkness, the section leaders led their heavily laden teams at a rapid pace. It was just past midnight when they arrived at a hill overlooking the radar site; they could hear the generator humming steadily. In the distance the sounds of the aerial bombardment could still be heard, but their intensity was lessening.

Suddenly, two automobiles came into view on the road, their headlights clearly visible in the dark. Tension rose – the road was only a few hundred meters from the sand dunes behind which the para-

NAHAL paratroopers cautiously approaching the radar site. Note the special heavy gear carried by the men.

troopers were crouching. The Egyptian cars stopped about two kilometers from the paratroop roadblock and doused their lights – nothing more was heard.

On a signal from their commander, the assault team quickly crawled forward, towards the sleepy, bored guards, fully visible as they advanced. In fact, one of the guards nearly stepped on an Israeli paratrooper!

The force was inside the radar site before the Egyptians realized it; one guard, crying out in terror as he faced the Israelis who had previously blackened their faces, gave the alarm and all hell broke loose as the paratroopers stormed the site, their automatic rifles blazing at anything that moved. In three minutes the position was secure; one section stormed the bunker, slamming hand gre-

nades into the ventilation slits. Several Egyptian guards were killed in the battle, with others, among them the site commander, escaping into the darkness. As the assault team mopped up the guards, the other team made for the dugouts, assembling their equipment to dismantle the radar shelters from their ZIL trucks.

As suddenly as it had started, the battle was over; all became silent once more as the dismantling team lit their acetylene torches with a hiss and started to work. The team had the shelters clear in less than an hour, while their comrades kept an alert eye out for intruders. But no one came – all remained silent as the work proceeded without interruption.

On the other side of the Gulf, in the forward command post, the tension became almost unbearable as the time passed. Officers, their ears glued to loudspeakers connected to radio sets, waited impatiently for word to come.

The dismantling team worked hard. Once the operations shelter was clear, many hands lifted its heavy weight in unison to make sure that nothing would hamper the lift; others, climbing like monkeys on the heavy Yagi antennae mast, pried the sections loose as workers below sawed away at the base.

By 0230 the shelters were ready to be lifted and the CH-53s were called in. The first chopper arrived overhead soon after, signalled in by an expert guide who positioned himself on top of the shelter, holding on to the rigging. The helicopter leader, hovering overhead, made a quick and perfect contact, and as he hitched the rigging to the electric hook, the loader jumped off. The chopper lifted the shelter upwards with a slight jerk, and off he went into the dark. As he cleared, the second CH-53 approached to lift the antennae shelter. Weighing a ton less than the other load, it lifted up much more easily and was away in minutes. As soon as the

helicopters' noise faded away to the east, the paratroopers laid demolition charges on all the remaining equipment and blew up the site, after which they withdrew rapidly to reach the rendezvous point where the Superfrelon helicopters were to pick them up. With their four Egyptian prisoners, they embarked quickly and were off before four o'clock in the morning. Egyptian reinforcements did not seem to arrive in the operations zone at all.

While the assault force was withdrawing, a last drama unfolded in the air. The captain of the lead CH-53 was an experienced helicopter pilot, intimately familiar with his aircraft. As he started lifting his heavy load, he watched the loading weight gauge tensely; the pointer quickly passed the 3,500 kg mark and went past the 4000 kg point. The pilot breathed a sigh of relief when it finally stopped at 4300. This was almost a ton more than the briefing weight figure. Without a moment to lose, the captain started to lift off – the weight held as the helicopter, engines straining, took height and veered off from the dugout in a cloud of dust. 'First shelter lifted', he reported over the radio, his message bringing a loud shout of joy from the senior commanders on the far shore. As soon as he gained altitude, a red warning lamp started winking on the control console in the cockpit. The heavy chopper shuddered, weaving alarmingly. Soon another warning light came on – indicating a fall in hydraulic pressure in the No. 2 pump. The main pump still seemed serviceable and the pilot, weighing danger against his chances of reaching the far shore, decided to put his trust in the aircraft and carried on with his precious cargo. His gamble paid off, and he set his shelter down on the beach to the loud cheers of the officers assembled in the cold dawn. Nearby, the second Sikorsky carefully landed the antennae shelter. Mission completed.

The Chief of Staff, filled with emotion, spoke to the assembled paratroopers and airmen: 'Your mission could well have been taken from a science fiction movie – only it was much more realistic'.

When the secret leaked out, the world's press exploded with vivid descriptions of the feat of arms. However, the real thing was much more fantastic than fantasy itself.

1973 YOM KIPPUR WAR
HOLOCAUST AT THE 'CHINESE FARM'

Yom Kippur, 6 October, 1973. Within hours after fierce fighting erupted in the Golan Heights and the Sinai Desert, the IDF Parachute Corps was rushed into action. One reserve brigade, commanded by Brigadier General Danny – now in his twentieth year of paratroop fame and still wearing his trademark, a full beard – took up defending positions in the Mitla and Jiddi Passes, well known to Danny and his men since 1956. Should the enemy decide to break through into eastern Sinai, it would be up to Danny's brigade to stop them. A regular brigade was flown to the Gulf of Suez, digging in to block Egyptian armor rushing south toward the Abu Rodeis oil fields. A third force was dispatched to Mount Hermon, now besieged by the Syrian Army.

As the fighting in Sinai progressed, Major General Ariel Sharon, commanding a reserve division in charge of the central sector, summoned Danny to his forward headquarters. Danny's paratroops, Sharon informed him, would be the first to cross the Suez Canal at Deversoir. The reserve brigade, which had fought on Ammunition

Danny Matt, as Colonel before attaining rank of Brigadier, IDF.

Hill during the Six Day War, was made up of veterans; many of them had been comrades-in-arms since the days of Unit 101. Despite their age – considered advanced by paratroop standards – the men in Danny's unit were still in top condition. The brigade was a first-class fighting unit, and Ariel Sharon knew it.

Paras advancing on the west bank of the Suez.

Now Danny listened attentively to his orders. He was to move westward with his men; a supply unit, carrying the 60 rubber assault boats needed for the crossing, would reconnoiter with the brigade somewhere on the road to the Canal. Danny, well accustomed to the promises of the Supply Corps, decided to organise his equipment on his own. Delegating an officer to procure all the necessary equipment – halftracks, jeeps, and even the rubber boats – Danny himself began to assemble his troops for the march westward. Ignoring the vehement protests of other units, Danny's 'requisitions officer' appropriated all the items on his 'shopping list', rejoining the brigade as the column was ready to move out.

The convoy inched its way forward, squeezing in between the already congested columns of tanks and supply trucks clogging the narrow desert tracks. Vehicles became bogged down in the sands, adding confusion to the already impossible traffic jam stretch-ing on into the night. Danny urged his frus-trated troops on, knowing the importance of attaining the objective as soon as possible. Incredibly, the brigade reached the road run-ning parallel to the Canal – only a few kilome-ters from the water – practically unscathed.

As Danny's force prepared to make its way down to the waterway, pandemonium broke out. Hundreds of Egyptian guns opened fire, showering missiles and artillery shells on the intersection. Keeping his nerve, Danny ordered his men to turn off the main road and proceed westward straight to a former IDF stronghold on the east bank of the Canal, now abandoned by both sides. The fact that the escorting armor had come under heavy attack – leaving many tanks burning around the intersection – did not prevent Danny from establishing his force in what was later called The Yard, a compound near the stronghold. Ordering the brigade artillery to concentrate its fire on the Egyptians holding the west bank, Danny lowered his rubber assault boats into the water. The lead team was on its way to Africa.

Amazingly, the assault platoons reached the west bank without incident. Clearing the barbed wire fences, the paratroopers quickly took up positions and dug in, signaling the rest of the brigade to quickly cross the water. By dawn of the following day – 16 October, 1973 – Danny's brigade, plus seven tanks fer-ried over during the night on rafts, was firmly entrenched in a narrow enclave on the west bank. The fierce battles raging on the eastern shore as Sharon vainly tried to shift his bridg-ing equipment to the crossing area, did not affect the small, exposed force on the Egyp-tian side. The persistent paratroopers now waited impatiently for Sharon's armor to catch up with them.

This, however, would be far from easy. By now, all the approach roads to the Canal had been blocked by Egyptian forces. Deter-mined armor attacks battered the Yard incessantly; equally determined, the Israelis defended the position as best they could. The IDF armor, despite repeated attempts throughout the night, had failed to bring the heavy engineering equipment to Deversoir; to make matters worse, it had suffered heavy casualties. After agonizing deliberations,

IDF infantry clearing the road to Budapest, mined by the Egyptians.

IDF troops advancing towards the Suez Canal.

Sharon decided to bring in the regular paratroop brigade from the Gulf of Suez to the 'Chinese Farm', a former Japanese-run agricultural station on the Canal approach road, now occupied by elements of the Egyptian 16th Infantry and 21st Armored Divisions. The paratroops, Sharon reasoned with a heavy heart, were perhaps the only forces that could break through the enemy concentration which had been harassing the Israelis incessantly for the past two days and nights.

As the paratroopers stepped out of their vehicles in the dark and adjusted themselves to the new terrain, their officers assembled the men by companies, readying them to move out. They had advanced only a few hundred meters westward when they were hit from all sides by a tremendous volley of fire. Confusion broke out as the brigade vainly tried to locate the source of fire. Fanning out as much as possible, the pinned-down paratroopers stormed the first Egyptian position – only to discover, to their consternation, that they had barged right into the divisional compound!

Crawling for cover and barely managing to dig in, the paratroops gasped as the whole area was suddenly lit up by volleys of flares, turning the darkness into full daylight. Tremendous artillery barrages – described by one survivor as 'missiles in machine-gun rhythm' – pinned the unit down. Many officers, endeavoring to lead their men into some semblance of order, were killed or wounded, leaving their troops leaderless at this most crucial time. Just when it seemed that nothing worse could possibly happen, the Egyptians began moving in their armor, firing hundreds of Guryonov heavy machine guns. Never before had the beleaguered paratroopers – despite years of combat experience on three fronts – seen such fierce fighting.

As dawn came, the survivors rallied and made one last attempt to storm the Egyptians. Despite their exhaustion and many casualties, they managed to establish a tenuous hold, but the firing did not let up for an instant. Finally, an Israeli tank unit came onto the scene, pushing into the compound to extricate the stricken paratroopers at all costs. Yet even this did not seem to daunt the Egyptians, and the battle raged on for hours more. At last, the combined paratroop and tank force succeeded in repulsing the exhausted Egyptians and breaching a narrow corridor through which Sharon could move his bridges. The paratroops had lost 45 men; nearly half of the unit had been wounded. But as they sprawled exhausted beside the road and watched the endless procession of IDF soldiers and materiel moving west to cross the canal, they knew there had been no choice.

Meanwhile, Danny's beachhead on the west bank had been discovered by the Egyptians. The brigade was rapidly subjected to all-out attacks lasting for days on end. Artillery barrages from some 400 guns, as well as air attacks, helicopter assaults, and commando raids, descended upon the reserve brigade. But the paratroopers held on. Finally, the IDF armor reached the west bank. Joined by the weary but undaunted paratroops, they fought fiercely in the heavy undergrowth, using each success to best advantage until the ceasefire left the Israelis firmly established in Africa.

Hundreds of kilometers away, another aspect of the war was being fought . In the first hours of battle, Syrian commandos had captured a vital Israeli observation post on snow-covered Mount Hermon. Now, as a ceasefire was imminent, the IDF had to retake the position; otherwise, it would be declared as belonging to Syria.

First contact with Budapest Stronghold.

THE BATTLE FOR MOUNT HERMON

Mount Hermon, which straddles Syria, Lebanon and Israel, is the highest point in the Middle East with its peak reaching 2814 meters. In the Six Day War part of the mountain's southwest crests had been secured by the IDF in a surprise move, heliporting Golani infantry onto the lower slopes and moving on by foot to the top. The Syrians, however, remained in control of the summit, which was some 600 meters higher than the position captured by the IDF, an omission which was to have a crucial bearing in the opening battles of the Yom Kippur War six years later.

Since Mount Hermon completely dominates the northern Golan Heights area, the control of its three peaks was of vital importance, furnishing distinct observation over a vast area covering the Damascus plain, Jordan and much of Lebanon. The IDF had therefore constructed an extensive observation complex in a well-fortified position. Fitted with highly sensitive electronic equipment, its radars covered much of the Syrian and Jordanian airspace, providing excellent early warning facilities.

The upper system of fortifications had not yet been completed and, as the position was manned by 60 men – mostly technical and administrative personnel from various services – it was badly maintained and signs of neglect were evident. Defended by only a handful of combat troops and commanded by a junior officer with 13 men at his disposal, it stood little chance of surviving a determined attack. The fort itself was well constructed to withstand massive artillery and air bombardment, using basalt rocks hewn out of the mountainside. A complex underground bunker system was designed to protect the sensitive equipment. But the surrounding trench and defensive position was incomplete and in bad shape.

The men manning the observation equipment had closely followed the Syrian build-up on the plain below, reporting continuously to their headquarters on the Syrian

IAF CH-53D helicopters carrying troops to the Hermon.

progress. Shortly after 1346 the artillery observation officer reported the Syrian gunners uncovering the camouflage nets from their guns; only minutes after, the Hermon fort came under heavy fire – the first shells exploding sent the men scrambling for cover. Soon after Syrian MiGs screamed over the mountain top, bombing and strafing as they thundered over the fort at low level.

The Syrians had originally planned to launch their assault on Mount Hermon at exactly 'H' Hour, but a delay in briefing had postponed the take-off of their heliborne force. The assault teams were selected from the crack 82nd Parachute Battalion and a ranger group, specially trained for this mission on a mock-up constructed by order of Brigadier General Bitar, director of Syrian Army Intelligence.

As the 'Syrian Hermon' position was fully visible from the Israeli fort, the Syrians had to mask the IDF position by fire before storming it. While the Israeli defenders raced for cover, the Syrians started their attack with an advance by a commando battalion of rangers jumping off from the northern hill and descending into the valley below the towering Israeli fort.

At 1445 the ranger group reached a point some 200 meters from the outer perimeter defense, coming under sporadic fire from the defenders who had followed their climb. At 1455 three Syrian Mi-8 helicopters landed near the upper skilift station below the fort, disgorging assault elements from the 82nd Parachute Battalion. One helicopter was hit and exploded, crashing in flames into the mountainside. The paratroopers took positions and started to exchange fire with the Israeli outposts nearby. Meanwhile the rangers closed in, supported by artillery fire and a last airstrike, then stormed into the inner courtyard while the surviving defenders retreated into the fort. Two heavy machine-guns were knocked out and the defenders were left with submachine-guns and hand grenades.

Just after three o'clock, the rangers and paratroops had linked up and were fighting in the fort. The remaining IDF defenders, however, put up a stubborn fight from the fort

The Hermon, captured by Syrian commandos, still smoking, as seen on 7 October from the upper ski lift station.

Golani troops advance from the upper ski lift toward the bunkers.

The Syrian positions are seen in the background.

itself, forcing the Syrians to withdraw.

At 1700 the Syrians attacked again, charging from the west with the setting sun at their backs. Scaling the walls of the fort with special grappling hooks and ropes, they forced their way into the inner fort, where fierce hand-to-hand fighting raged. The Syrians pushed their way by sheer weight of numbers, clearing underground passages, until, reaching the main communication and monitoring center, they blew in the last protective steel doors. The fort, with its remaining survivors, fell after a courageous battle.

On the morning of 7th October, Golani infantry mounted a counter-attack aimed at recapturing the fort. Moving up the winding road they were ambushed by Syrian commandos and forced to withdraw, losing two dozen dead.

The Syrian ranger group remained in all-round defense on the mountain top, bringing in heavy equipment in anticipation of a new Israeli attack – well aware that the IDF could not ignore the vital importance of this position.

During the following days, as the savage battles raged on the plains below, the fort remained relatively unmolested, apart from an occasional air attack by IAF aircraft. However, many important visitors came to inspect the electronic equipment, which was later dismantled by Soviet experts.

The IDF was already some 30 kilometers inside Syrian territory, beating off the last combined Syrian and Iraqi attacks, as General Hofi, heading Northern Command, ordered the recapture of the Hermon, anticipating that time was running short before a ceasefire would freeze the positions and leave the Syrians with their prize intact.

The plan was to recapture the fort by a two-pronged attack made by Golanis and a paratroop force. The paratroopers were to storm the Syrians from above as Golani infantry climbed up from below. The attack was scheduled to start on 21 October.

At 1400 the paratroops, commanded by Lieutenant Colonel Hezi, were heliported to their positions. With F-4E Phantoms flying air cover, IAF CH-53Gs loaded with paratroopers flew over the northern slopes into

Israeli POWs.

Recapturing Mount Hermon.

Lebanon and landed northeast of the Syrian Hermon position. The IAF helicopters chose the northwesterly route which would shield them from Syrian radar and avoid the heavy SAM and anti-aircraft defenses in the Damascus plain to the east.

In order to accomplish safe landing on the high altitude ridges of Mount Hermon, each helicopter was loaded to only half its normal carrying capacity. To achieve surprise, no landing zone preparations were made and the choppers landed directly on the Syrian positions, unopposed. While the Phantoms continued their air support activities, the remaining elements of the paratroop force were shuttled to the IDF landing zone, landing slightly to the northeast in order to avoid interfering with the advancing assault teams, which were supported by organic mortar and artillery called in by observers from batteries deployed below. IAF attack aircraft flew interdiction missions against Syrian reinforcements coming up the mountain road from Qatana.

Shortly after 1600 the Syrians began to comprehend the threat to their vital positions on Mount Hermon and reacted violently, with repeated air attacks which were intercepted by IAF Phantoms, resulting in the shooting down of 11 aircraft. Some Syrian fighters flew at zero level, scraping the mountain ridges in an attempt to attack the paratroopers, but were chased by low-flying Phantoms through valleys and riverbeds and shot down into the mountainside.

As the air attack proved ineffective, the Syrian artillery started pounding the mountain top, attempting to heliport reinforcements, which were also intercepted by IAF fighters before they had a chance to land. By then night had fallen, but the battle for Mount Hermon was only beginning.

Under cover of darkness, the paratroopers stormed the Syrian serpentine position, situated at point 2814 – the highest on the mountain ridge. Working their way down to the south, they captured the Syrian command post without loss. By 0330 the whole Syrian Hermon position was secure.

Meanwhile, jumping off from Majdal Shams, the Golani infantry brigade moved up the mountain from the south on three routes, similar to the unfortunate attack a week earlier. Leading were five Centurion tanks. In almost the exact spot where their previous attack had broken down, the brigade vanguard was engaged by Syrian fire. It was a battalion-sized force of commandos, scattered over the rocky mountainside, which now blocked the advance. Using night telescopic sights and the advantage of height, the Syrians picked off the Israeli soldiers, who had difficulty in pinpointing the masses of targets in the dark from the road below. The tanks and APCs tried to move on, but were stopped by a roadblock and came under RPG fire which set one vehicle on fire.

As the infantry dismounted to advance on

The recapture of the Hermon fortress; a Centurion tank in the lead.

IDF helicopters bring paratroopers to the Syrian position overlooking the Israeli Hermon stronghold.

Golani, supported by three halftracks, advance toward the upper ski lift station.

foot, both the brigade commander and the lead battalion commander were wounded. The fighting became desperate, casualties mounting by the minute. Junior leaders took command and pushed forward, dragging the men with them up the mountain while they ducked the deadly fire from the Syrians above.

To ease the situation, two companies of Golani, in reserve below, were flown to the lower ski lift by helicopters, while the paratroopers on the Syrian Hermon were urged to move down. Slowly the Golani fighters, led by junior officers, fought their way up among the boulders, mopping up the Syrian commandos by sheer courage as they went.

All through the night the terrible fight went on, with single groups fighting independently, determined to reach the mountain top. As dawn broke, the brigade reconnaissance squadron captured the upper ski lift position, strongly defended by dug-in Syrian commandos. Now, near their objective, the Golani infantrymen, totally exhausted by the night's fighting, rallied for the final assault, with the brigade operations officer personally leading the attack, calling in close artillery support.

The Golani survivors recaptured the fort, and raised the Israeli flag on one of its antenna masts. The Syrians, equally exhausted, rose from behind rocks and boulders to give themselves up or tried to escape – only to be rounded up by the advancing paratroopers.

By 1000 hours on 22 October, shortly before the ceasefire went into effect, the Hermon position was secure in IDF hands.

But the cost was heavy: 55 Golani fighters were killed and over 100 wounded during the night. Many of the casualties were officers, among them the brigade commander, Brigadier Amir Drori*, and two battalion commanders, injured while leading up front.

One young sergeant, leaning exhausted on a rock, looked at the battered fort, hardly believing that he was alive. Sipping the tepid water from his canteen he said, 'We were told that Mount Hermon was the eyes of the State of Israel and we knew that we simply had to take it, whatever the cost.'

Golani infantry proved its worth again under the most trying circumstances, making it a truly elite fighting unit.

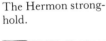
The Hermon stronghold.

*Later, as Major-general, head of Northern Command and overall commander of the Lebanon campaign.

BEAUFORT CASTLE – 1982

Beaufort Castle, built in the 12th century by the Crusaders, overlooks the Litani river gorge in Southern Lebanon. From 717 meters high, its straight wall dominates the river bed and the strategic Hardale bridge below. One hundred and twenty meters long and 60 meters wide, the medieval castle, built from local rock, was fortified by the PLO with concrete positions on the ramparts, and deep stone-walled trenches connecting well-situated bunkers, all rendering depth to the Beaufort's defense.

The high mountain wall discouraged invaders from the southeast, while access from the west along a steep, narrow road through the fortified village of Arnun was also difficult, the road being fully covered by fire from the fortress and village.

Beaufort Castle was turned into one of the most important observation posts overlooking the Free Lebanese enclave of Major Sa'ad Haddad, as well as the Israeli Galilee panhandle. From behind its stone walls, PLO and Syrian observers sent artillery shells and Katyushas raining down on civilian targets, making life miserable for both Israeli and Lebanese alike. The fort became a notorious target for the IDF and Haddad militia, but intense artillery bombardment and repeated air strikes had little effect on its defenders, who trusted their lives to the age-old rock enclave – a tribute to the genius of Crusader engineers.

Never yet overrun in battle, Beaufort Castle in June 1982 stood in silent defiance as Israeli troops smashed over the Lebanese border and their armored spearheads ground up the mountain roads and fanned out to the west. The castle was already under fire from artillery and air strikes, which seriously impeded its anti-tank guns and fire ob-servers, as the Sayeret Golani recce squadron, on M113 APCs, made its way forward. Crossing the bridge and climbing up the Arnun ridge, the Sayeret spearheads came under fire from units of the PLO Kastel brigade defending both the village and the fortress above it.

By now night was falling. The vanguard, made up of combat engineers, crashed through the village and raced up to the castle under intense fire. Following this small force, the Sayeret's APCs advanced. One vehicle was hit by an RPG and exploded, the unit commander severely wounded. Casualties mounted and the convoy stalled, with fire from the castle and village gaining strength.

Listening to the radio net from his CP

Beaufort Castle.

APC arriving at the Beaufort the morning after.

Infantry during the Peace for Galilee operation.

situated below, the brigade deputy dispatched a replacement leader up the mountain road. Major Guni Harnik, a former Sayeret commander, had earned a high reputation, and his deep, familiar voice calling 'This is Avenger, coming in' became an immediate morale booster to the attackers trying to extricate themselves from the chaos below the fire-spitting castle. On the way up the tortuous mountain road, Guni's APC overturned, but he escaped relatively unhurt. He took his Galil rifle, ran up the mountain, and joined his men. Quickly organizing the attack, Harnik directed the engineers to storm the southern part of the compound, with a fire base to the north. The recce unit itself was to assault the castle, its bunkers, and the high ground.

As the final charge began, the teams raced the last two hundred meters and disappeared into the trenches, firing from the hip and hurling grenades through the gun slits. The stone and concrete trenches turned out to be too narrow for the heavily loaded infantrymen. Working their way forward with difficulty and exposed to fire, junior officers raced over the open ground slamming explosive charges into gun ports that were spitting fire.

The assault teams advanced, silencing bunker after bunker. But PLO gunners were stubborn and fired back non-stop. Casualties were mounting by the minute. While combat medics attended the wounded, the survivors kept the assault fanning out to all sides, saturating the Arab defenders with offensive fire.

By now Guni had joined the forward group and was leading the attack on the main bunkers. This position was particularly well-situated, closed on all sides with firing slits close to the ground. With deadly machine gun fire being sprayed low over the boulders, Harnik and another officer raced forward and identified the openings. They hurled grenades and explosive charges into the bunker, until Guni was mortally hit.

The junior officer took command and directed his men in taking the remaining fortifications. Suddenly, dark shadows started to appear on the castle walls, and the Beaufort's PLO defenders broke and ran. By dawn the castle was in Israeli hands. On their antennae the Sayeret Golani soldiers raised their unit's flag, which was clearly seen from the valley below and back home across the border. The eagle's nest was secure, but Sayeret Golani – and Israel – had paid a heavy price.

Amphibious landing of
Israeli elite forces in
Lebanon.

Israeli troops entering
Lebanon.

99

Paras transported up
to the southern Golan
by Sikorsky s-58
helicopters; 1967.

Recon jeep in the
Sinai; Yom Kippur
War.

Relieving the besieged
Budapest stronghold;
1973.

IDF infantrymen in a
sandbag position west
of the Suez Canal.

ADEN 1963-1967

The Aden Emergency consisted of two distinct but simultaneous campaigns, one a tribal uprising in the Radfhan mountains, and the other a wave of urban terrorism in Aden itself. Both had their roots in Arab nationalism and were actively supported with funds and arms by President Nasser's Egypt. Both were also safely contained until February 1966, when Harold Wilson's Labour administration announced that the Aden base would be abandoned, and with it the Federal government which had been set up to control the colony and its adjacent sultanates. This decision promptly rendered any sort of counter-insurgency policy meaningless, and the only outstanding question was which terrorist group would rule after independence; in fact, the last year before the British withdrawal was marked by an outright civil war between the National Liberation Front (NLF) and the Front for the Liberation of Occupied South Yemen (FLOSY), whose hatred for each other surpassed anything they might have felt for their former colonial masters. Once the NLF had emerged victorious, this inept policy led to the creation of a Marxist republic, the creation of a Soviet naval base at the southern entrance to the Red Sea, and a temporary loss of British credibility throughout the area. From this seedy political debacle only the British armed services emerged with their honour intact, having performed all that was asked of them and beaten the enemy in the field.

The Radfhan mountains, a wilderness of red, heat-shattered rock where temperatures can rise to 130 degrees, are peopled by ungovernable tribes who regard war as sport and who go fully armed about their daily business. They are always ripe for trouble and in 1963 the twin spurs of fervent Arab nationalism and an ample supply of Egyptian arms coming across the border from Yemen made a major explosion a certainty. A combined British/Federal task force was sent in to restore order in January 1964 and then retired, believing its mission to have been

Britain's Irish Guards patrol in the notorious Aden Crater quarter.

102

completed. The tribes immediately took the withdrawal as a sign of weakness and promptly rose again. This time a strong brigade battlegroup was employed to quell the revolt and based itself in the Thumeir basin in the heart of the Radfhan, using artillery, airstrikes and armored cars to hammer the enemy. Among the hills themselves, however, it was the superbly fit elite forces, including 45 Commando, Royal Marines, a company of the Parachute Regiment and an SAS squadron, which beat the tribesmen at their own game. Nonetheless, the operation began badly when an SAS patrol, covertly directing air strikes, was detected and surrounded. The patrol fought back, inflicting sharp losses, and broke out the following night, twice ambushing its pursuers during the march back to base. Two of its members had been killed, however, and the tribesmen savagely decapitated their bodies and triumphantly exhibited the heads. The affair caused a storm of international revulsion and was counter-productive in that the British now fought with a sharp vindictive edge. One by one, the peaks dominating the cultivated areas of the Radfhan, which formed the heart of the revolt, were taken by the Commandos and Paras. The majority were captured by hard climbing and hard fighting but some were seized by *coup de main* following careful observation of the enemy's positions which

British SAS patrol in the Radfhan. Note the US M16 assault rifle used widely by SAS.

revealed that the tribesmen were as fond of their wives as they were of their Russian-made weapons, and would abandon their sangars after dark to visit them. The return of these dutiful husbands in the hour before dawn was marked by accurate bursts of fire which dropped them among the rocks. After five weeks the principal massif was in British hands and the dissidents gave up, the most troublesome elements being banished from the Radfhan for the duration of the emergency.

In Aden itself slow but steady progress was maintained against the urban terrorists by means of covert observation, road blocks, aggressive patrolling and cordon and search operations. The SAS also employed what were called 'Keeni-Meeni' techniques which involved plain clothes patrols by soldiers who could pass as Arabs. These patrols stalked their prey among the teeming alleys and put

an end to the careers of a number of Cairo-trained assassins. All this patient and productive work went for nothing with the announcement of the British withdrawal and terrorist activity escalated to unprecedented levels. There was, nonetheless, some cause for satisfaction in that many of the terrorist targets were rival terrorists. In Sheikh Othman, for example, the NLF and FLOSY fought a large scale gun battle in the best Wild West tradition, with screaming bodies tumbling from the rooftops; this was watched with some interest by the 1st Battalion The Parachute Regiment, who then moved in to clean out the dazed survivors.

The terrorists made no attempt to interfere with the final evacuation, and the troops left without the slightest regret. Aden, with its heat and smells, had always been an unpopular posting, and even at the best of times had been referred to as 'Arabia's arsehole.'

British soldiers in Aden with general view of the city and fortress.

Ferret armored car of the KRH patrolling in Aden.

Aden (Radfhan Mts) – infantry patrol is re-supplied by helicopter.

Royal Navy landing craft in Persian Gulf.

OMAN AND THE DHOFAR WAR

The northern territories of the Sultanate of Muscat and Oman, and notably the Musandam peninsula, lie at the entrance to the Persian Gulf, through which passes a high proportion of the free world's oil supplies, and for this reason the United Kingdom has maintained a traditional policy of military support for its rulers. The SAS has fought two campaigns in Oman, the first in 1958/59 against dissidents who wished to depose the Sultan and who had installed themselves on a large fertile plateau known as the Jebel Akhdar or Green Mountain. This towered 10,000 feet above the surrounding plain, with only twelve known and easily defended routes up the sheer cliffs which encircled it. Throughout history, this natural fortress had never been stormed, but in January 1959 an SAS squadron, covered by diversionary attacks and a deception plan, made an astonishing climb up an untried approach in total darkness, and reached the summit. A subsequent supply drop and the presence of British troops on the plateau convinced the dissidents that a full-scale airborne assault was under way, and the rebellion promptly collapsed.

The second and largely unreported campaign was fought between 1970 and 1976 in the Dhofar mountains, lying in the southwestern corner of the country, and resulted in the defeat of attempts by the Marxist Republic of South Yemen (formerly Aden) to destabilize Oman by guerrilla activity. The most significant action of this war took place on 18 July 1972 at the town of Mirbat, which was held by a 10-strong SAS training team under 23-year-old Captain Michael Kealy, 30 Omani Askaris and 25 Dhofar Gendarmes, the last mentioned being armed with obsolete weapons. At dawn the little garrison was attacked by a force estimated to number 250, armed with AK-47 assault rifles, mortars, machine guns and anti-tank rockets. The battle raged at close quarters for three hours with the utmost fury. Kealy had, however, managed to radio for assistance and shortly after 0900 Strikemasters of the Sultan's Air Force bombed and strafed the invaders while SAS reinforcements arrived by helicopter. The attackers fled, leaving many of their casualties behind; their precise loss is not known but in the battle itself and the violent recriminations which followed it is believed that about half were killed or wounded. No further attacks of this nature were ever attempted. The garrison's casualties amounted to four killed (two of them SAS) and three wounded, including one SAS soldier. Kealy, who had been the soul of the defense, was awarded the DSO; tragically, he was to lose his life in 1979 while taking part in an exercise in the Brecon Beacons.

British Forces in the
Radfhan Mountains.
Photos above show
machine gun position,
Saladin armored cars
on combat patrol, and
a typical view of Brit-
ish forward positions.
Left shows a Saladin in
action, and a British
ambush at Thumeir.

THE AMERICAN RESCUE MISSION TO IRAN

US Navy helicopters prepared for flight before the rescue attempt.

US Navy helicopters, specially painted for the BLUE LIGHT operation, taking off from the Nimitz.

Although its failure made the headlines around the world, the American's abortive mission nevertheless included heroic action by its participants. Had it succeeded, it could have changed many issues in the region. It was the first time that American forces were ordered into action in the Persian Gulf area.

Taking off from USS *Nimitz,* the RH-53 Sea Stallion helicopters met at the Desert 1 rendezvous with the C-130 Hercules flying combat personnel and equipment from Cairo West and Kena in Egypt. The whole operation was directed from AWACS cruising in the area acting as a central command post, enabling direct control of the field of action by link to the Pentagon and the President. This marvellous feat of modern communications may well have been the

undoing of the mission. It can be argued that, had the commander on the spot been left to make his own decisions, he may well have decided to continue the mission in spite of its mishaps.

The helicopters prepared for the operation.

ALGERIA 1954-1962

No sooner had the regiments of the Foreign Legion returned from Indo-China than they found themselves bearing the brunt of the long and bitter Algerian War of Independence. This was complicated by the presence of French colonial settlers, many of whom had lived on the land they farmed for generations and who were violently opposed to the prospect of being ruled by the Moslem majority. Thus, the normal pattern of guerrilla warfare was aggravated by civil atrocities, the net result of which was to harden the attitude of the *Front de Liberation Nationale* (FLN) and make a moderate compromise solution less acceptable to both sides. After the Suez operation of 1956 the Arab world in general turned against France and the FLN received all the financial and military support from abroad that it needed. Despite this, the FLN suffered repeated defeats in the field and most of its leaders were killed or captured. By 1959 the military situation had been brought under control. At this point President de Gaulle, aware that the French public was sickened by the brutal nature of the war and tired of paying for it, announced that Algeria was to be granted self-determination. To the Army and the colonists this seemed like betrayal and anti-Gaullist feeling within the ranks built up to the point when, in April 1961, several generals, with

Soviet air support of ground forces, with Mi-24A Hind helicopter.

support from their parachute troops, seized Algiers and held it against the government for five days. The coup collapsed when French public opinion revealed itself to be solidly behind de Gaulle. The Legion, which preferred an honest soldier's war in the mountains to politics, stood apart, with the exception of the *1e Régiment Etrangère Parachutiste*, which was disbanded for its part in the mutiny. By the end of 1963 the Legion had left its historic depot at Sidi-bel-Abbes forever and its units were scattered between France, Corsica, Djibouti and Madagascar.

AFGHANISTAN

On 24 December 1979 the Soviet Union launched an invasion of Afghanistan in order to prop up the tottering communist regime in Kabul. The invasion was spearheaded by three airborne divisions, one of which, the 105th, secured Kabul airport, while two more, the 103rd and 104th, occupied the Bag-

ram airbase to the north; simultaneously, motor rifle divisions rolled across the frontier to seize the provincial towns and effect a junction with the airborne troops. By January 1980, some 50,000 Soviet troops had secured the main administrative centers throughout the country, were in control of the Khyber and Khojak Passes, and had moved the 40th Army Headquarters into a permanent location at Bagram. As the build-up continued, further units advanced as far as the Iranian border, both in the central region and in the desert area to the south.

The Soviet invasion met little opposition, but the Russians were soon to discover a truth that should have been evident from the briefest study of the region's military history – namely, that it has always been deceptively easy to overrun this country, but almost impossible to maintain oneself there, for whatever internal differences the Afghans may have, they have always put these aside to engage an invader with their own brand of ruthless and efficient mountain guerrilla warfare. It quickly became apparent that the

Soviet soldier in Afghanistan. The Soviets deployed their elite airborne divisions as well as several motorized formations to Afghanistan. This solder is armed with an AK-47, and is wearing winter fatigues.

Russians were taking quite heavy casualties and that in reality they controlled only the towns and their fortified garrisons, and that movement between these was only possible in heavily escorted convoys. Further, their Moslem Turkoman and Uzbek motor rifle troops were less than enthusiastic about making war on their neighbors and co-religionists, and have had to be replaced by European formations drawn from the Soviet Army's Central Reserve.

Thus far, despite their helicopter gunships, tanks and artillery, the Russians show no sign of bringing the situation under control, while the guerrillas seem to grow stronger and bolder as they capture more and more weapons. Again, there is no evidence to suggest that the elite Airborne and Airborne Assault formations have fared any better than the rest. Equally, any British veteran of the North-West Frontier will confirm that the Soviets' inexperienced SPETSNAZ (Special Forces) troops can best be regarded as amateurs in clandestine warfare when it comes to fighting the tribes in their own mountains, and risk the type of death only an Afghan can inflict, to boot. The Kremlin seems to have created a problem for itself which its Army cannot solve.

Soviet 105th Airborne Division BMDs in Kabul.

AFRICA
THE ENTEBBE RAID

For the sake of security, many brave deeds of Israel's airborne forces must remain forever untold. One incredible action, though top secret throughout its feverish planning, may now be published in great detail. The victory scored by the IDF paratroops and commandos at Entebbe has gone down in history as the most incredible raid ever accomplished by any army in the world.

The story of Entebbe began on 29 June, 1976. Air France Flight 139, enroute from Tel Aviv to Paris via Athens, was hijacked by terrorists operating under the aegis of the Palestine Liberation Organisation. After landing at Benghazi, Libya to refuel, the pilot of the French Airbus was directed to head for Uganda.

Upon landing late that evening at Uganda's Entebbe International Airport, the passengers were taken under heavy guard to the airport's old terminal. During the seemingly endless week spent as 'guests' of smiling dictator Idi Amin, the non-Jewish hostages were released and flown to safety. The Jews, however – along with the courageous crew of the Airbus who elected to stay with them – were not to be released unless the Government of Israel decided to accede to the terrorists' demands. These, including the liberation of 54 convicted terrorists, were unthinkable. But so was the murder of the hostages. Out of the seemingly insoluble dilemma arose a third alternative – to free the hostages by military action.

The choice, to any other government, would have been impossible. Entebbe, almost 5000 kilometers from Israel, was guarded by the Ugandan Army, not to mention the terrorists themselves. But Chief of Staff Motta Gur had confidence in his men. Gur delegated the planning of the land forces' role in the action to a young and energetic officer, Brigadier General Dan Shomron. Shomron, a veteran of many bloody battles and now o/c Paratroop and Infantry Forces, was assisted by the o/c Israel Air Force, Major General Benjamin Peled, whose job it was to chart the mission of the IAF. The final plan – approved by Israel's Cabinet at the last possible moment – involved a handpicked team of paratroops and commandos to be flown out to Entebbe in four C-130 Hercules transport aircraft. Once the aircraft had landed on the runway, the troops were to rush out, storm the old terminal, dispose of the guards and terrorists, and escort the hostages to the Hercules waiting to fly them home.

The leader of the team, Lt. Col. Jonathan 'Yoni' Netanyahu, was a truly incredible character. This odd mixture of intellect and guts was brought up in New York and Jerusalem by Israeli parents. On reaching the age of 18, Yoni returned to Israel to join one of the elite units of the IDF. Rising quickly in rank, he distingu-

Facing page:
Artist's rendition of the assault at Entebbe (Champion, UK, 1979).

Lt. Col. Yonatan Netanyahu.

uished himself in the Six Day War of 1967, where he was seriously wounded and discharged from the IDF with a 30% disability. Less than two years later, still 30% disabled after a series of operations, he returned to the army and, despite his wounds, was given command of a crack commando force. He advanced in rank, participated in a series of audacious raids, and in the Yom Kippur war rescued one of his friends, an armor battalion commander, wounded and left to die by the Syrians. Then, Yoni commanded a tank battalion until late 1975, when he returned to lead an elite unit. Now, on 4 July, 1976, his body squeezed into the front seat of a black Mercedes lashed down inside the lead Hercules, Yoni faced the toughest mission of his life.

As the first Hercules, piloted by the transport squadron leader, rumbled down the runway, Yoni's men settled into their bucket seats, preparing for the long journey ahead. Their commander had trained them to the utmost; they had sprinted down a training course matching the distance from the aircraft to the Entebbe terminal time and time again, until they could finish the sprint in less than 120 seconds. Now, hanging by its turboprops, the lead aircraft took to the air and headed south, followed closely by the other three transport planes.

As they flew down the Gulf of Eilat, the four Hercules took necessary evasive action in an effort to dodge the watchful radars of Egypt and Saudi Arabia. Once over the Red Sea they set course on a route which would take them deep into Africa. The lead pilot, listening in on the radio, tensed as he heard the latest weather report – clouds and thunderstorms over the African airspace. This could ruin the timetable, scotch the entire mission, he thought. Then, glancing back into the cabin, he saw the calm faces of the paratroopers, and was himself reassured.

Thick clouds covered the sky and driving rain beat down on the windshields; the monotonous sweep of the windshield wipers alternated with the fitful pitching of the airframes in the turbulence. Unearthly flashes of lightning illuminated the black sky; rolls of thunder vied with the noise of the engines. The pilots, fighting the controls, kept the aircraft in strict formation. As the Hercules approached their target, Yoni slithered out of the Mercedes and joined the lead pilot for a moment in the cockpit. A reassuring hand on the shoulder, a confident wink, and the young commander returned to his station, readying his men for action.

Over Lake Victoria, the formation split up, each pilot taking his station for landing according to plan. Amazingly, Entebbe Airport was fully lit, runways ablaze with highly visible landing lights. The lead aircraft turned warily into the final leg of the approach, glided silently onto the runway, and stopped precisely on the spot called for by IAF intelligence. Incredibly, the landing was just 30 seconds late – an achievement rarely matched even by veteran airline pilots who had flown the Africa run for years.

Yoni and his men in the black Mercedes rolled down the lowered ramp of the first aircraft, followed by two unmarked Land Rovers. Scant seconds behind them, paratroops began pouring from the other aircraft. The black Mercedes rolled past an astonished Ugandan sentry, who, unable to perceive the whiteness of the driver's face, mistook him for a senior Ugandan officer; saluting, he stepped back to let Yoni through. Now the assault parties had a clear path to the terminal. Surprising the terrorists before they could shoot, the Israelis gunned them down one by one. Then, shouting for the terrified hostages to lie low, the airborne forces stormed into the terminal, submachine guns blazing. Support groups silenced the Ugandan guards, who had begun to concentrate fire on Yoni's men. A bazooka smashed the control tower searchlight; under cover of the resulting darkness, the paratroops began to assemble the hostages for the journey home.

Suddenly, horrifyingly, a lone shot rang out in the darkness – and Yoni Netanyahu toppled forward. Swiftly, his second-in-command took over; within minutes, the hostages were inside the aircraft, ready to take off. As the heavily laden transports rolled down the runway, the entire airport was suddenly illuminated by a vast display of fireworks. A special demolition team had blown up the MiG-21 aircraft constituting Uganda's fighter squadron. Now the Israeli aircraft were safe from pursuit.

Hours later, early on the morning of 4 July, 1976, four IAF Hercules aircraft flew over the towns and cities of southern Israel to land at Ben Gurion Airport. The rejoicing – unparalleled in Israel's history – grew even wilder as a tide of excited Israelis literally swept the intrepid soldiers off their feet. Crowds danced horas on the tarmac; flowers and champagne were showered on the victors. It seemed as if the whole state were one large carnival. But this was not quite the case. Inconspicuously and with incongruous silence, a stern, sad band of tired men filed out a side exit. On their shoulders, in an unwitting parallel to the red-bereted troops being carried shoulder-high by the jubilant crowd nearby, they bore the lifeless body of their commander, Yoni Netanyahu – the only man lost in the raid.

The first Hercules brings the hostages back to Lod.

The terminal at Entebbe.

CENTRAL AND SOUTHERN AFRICA

The past forty years have witnessed numerous limited wars in Black Africa, some of which have been wars of independence, some civil wars, and a comparatively few wars between states. The involvement of elite forces in these conflicts was unusual, but by no means unknown. The Foreign Legion, for example, has twice intervened in the civil war in Chad, and has also put down a guerrilla movement in Djibouti.

nized transport could not penetrate – a technique also used with some success by the Portuguese Army against its opponents in Angola and Mozambique.

The South African Army, which includes one parachute brigade in its order of battle, is the most efficient on the continent and has carried out a number of damaging punitive strikes against guerrillas operating from bases in Namibia, Angola and Mozambique.

These South African soldiers, seen during a cross-border raid into Angola, are fully equipped for long range operations; note the unique 'double-sided' backpack on the soldier in front, with PRC 25 communications set on his chest and rucksack on his back.

During the struggle which led to the creation of Zimbabwe the Rhodesian Army possessed several Special Forces units which were highly effective. These included three locally raised SAS squadrons; the Selous Scouts, the majority of whom were black soldiers whose function was covert surveillance and tracking guerrilla bands in the bush; and Grey's Scouts, a mixed cavalry unit used to pursue guerrillas in country which mecha-

Several other African states have raised 'parachute' or 'commando' units, but whether these approach the high standard of training required to justify such titles is a matter for debate. When in trouble, the tendency has been for Black African armies to look for outside assistance, raising their own 'instant elites' in the form of white mercenary units. These usually contain a high proportion of former soldiers, some of whom have served in genuine elites, but much depends on the standards of leadership and discipline imposed, as well as a degree of selectivity in recruitment. On the one hand those mercenary units involved in the Congo (Zaire) debacle of 1960-1967 performed a valuable military role and saved many innocent lives; on the other, those mercenaries recruited to fight in the Angolan civil war gave the impression of being mindless psychopaths incapable of performing the simplest military task. States which are clients of the communist bloc tend to recruit units of the Cuban Army, which appear efficient in the purely local context but have performed poorly against first class enemies such as the South Africans. The Soviet Union is itself directly involved in the chaotic affairs of Ethiopia and in 1974 assisted in repelling the Somali invasion of the Ogaden region. Subsequently, Soviet firepower and helicopter gunships have been used against separatist rebels in the province of Eritrea, with as little result as in Afghanistan.

Once, the shifting sands of African politics encouraged international terrorists to believe that they would find a secure refuge in the Dark Continent. In 1976 this belief was rudely shattered at Entebbe, Uganda, then ruled by the unbalanced Idi Amin, when an Israeli airborne commando flew in to rescue

the passengers and crew of a hi-jacked French airliner, who were guarded not only by the terrorists but by Ugandan troops as well. The mission succeeded brilliantly, much of Amin's Air Force being simultaneously destroyed on the ground to prevent pursuit. The following year a German GSG-9 anti-terrorist unit, with British SAS assistance, stormed a hi-jacked Lufthansa jet at Mogadishu, Somalia, wiping out the terrorist group responsible. The lesson of these two operations was that there are few areas of the world that Special Forces cannot reach, and there was an immediate decline in similar terrorist activity.

South African patrol in Namibia. Their equipment consists of FN 7.62 rifles and light combat webbing.

South African infantry in training. Note the Savannah camouflage.

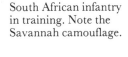

THE SOUTH ATLANTIC 1982

The Argentine invasion of the Falkland Islands and South Georgia was led by the *Buzo Tactico*, a unit which styled itself on the SAS, and by the Naval Infantry Corps, which had the reputation of being the best in its country's service. In each case the opposition consisted of a tiny Royal Marine garrison, armed with nothing heavier than an 84mm anti-tank weapon, but both put up a tremendous fight until overwhelmed by sheer weight of numbers. The total Argentine loss was never admitted but probably amounted to about thirty killed and a higher number wounded, three members of the *Buzo Tactico* captured, an APC knocked out, two helicopters shot down and a frigate very seriously damaged; incredibly, there were no fatalities among the Royal Marines and only one man sustained serious injury. Shamefaced, the Argentines were later to claim that for humanitarian reasons they had not opened fire, but the facts were against them.

The Argentine garrison on South Georgia surrendered to an SAS detachment on 25 April without firing a shot in its own defense, and from 1 May onwards SAS and Special

British landing craft carrying troops to shore. Note in the background the Type 12 Frigate and the QE2 luxury liner pressed into troop transport service for the Falklands operation.

Argentine forces land in the Falklands with their LTVP-7s.

Argentine marine commandos practice shore landing during exercises.

Boat Squadron (SBS) teams were being inserted into the Falkland Islands themselves. Ten such patrols are known to have operated on West Falkland, twelve on East Falkland and four on various off-shore islands; in five cases there was physical contact with the enemy. The information gathered revealed the nature of the Argentine dispositions and led directly to the successful landing at San Carlos. The SAS raid on Pebble Island on 14 May destroyed six Pucaras and several other aircraft, thereby eliminating a large proportion of the enemy's ground-attack force. There was, too, the strange incident of the British troop-carrying helicopter which landed in Chile, having flown close to or over Argentine territory. Shortly after-

wards, reports from Chile claimed that dispersal at air bases in southern Argentina was being practised to an unheard-of degree, and since British air activity did not extend to these bases, the conclusion is obvious. It has also been suggested that SAS and SBS teams operating near these bases provided early warning of Argentine sorties against the British Task Force.

Once the San Carlos beach-head had been consolidated, the British began their epic and heavily-laden march across country to Port Stanley, led by 42 and 45 Royal Marine Commandos and the 3rd Battalion The Parachute Regiment. In the opinion of Brigadier-General Menendez, the Argentine commander, the intervening waste of bog, heath and wind-

The submarine *Santa Fe* settles at the Grytviken whaling berth; 28 April 1982. It took three direct hits, one of them passing straight through her conning tower (note the hole) without exploding. A Wasp helicopter from the HMS *Endurance* hovers above the vessel.

British paratroopers storm ashore to secure a beachhead on East Falkland Island; 21 May.

A Royal Fleet Auxiliary logistic landing ship with the Falklands task force on its way to the islands. Later in the campaign, these vessels were attacked at Bluff Cove, resulting in the destruction of the *Sir Gallahad* and *Sir Tristram* with heavy loss of life.

swept upland was impassable; in fact, these were the very conditions in which Marines and Paras alike had completed their initial training on Dartmoor and the Brecon Beacons. The flank of the march was covered by the 2nd Battalion The Parachute Regiment which, on 28 May, won the battle of Goose Green against odds of 3:1, inflicting ten times their own loss. Three days later, in response to an SAS report that Mount Kent, the highest of the hills overlooking Port Stanley, was weakly held, a Commando company was lifted in and secured this vital position. As the march continued a unit from the *Buzo Tactico* was located in Top Malo House and engaged by an equal number of the Royal Marines' Mountain and Arctic Warfare Cadre. After the Commandos had blown off the roof with anti-tank missiles, which also

British airborne troops ready to disembark from their transports to landing craft off the Falklands coast.

The Royal Marines ashore.

Royal Marine Commando firing 81mm mortar on Ascension Island while waiting for ships to take them to the Falklands.

set the building ablaze, the Argentines came out fighting; four of them were killed and seven wounded, at a cost to the Marines of three wounded.

The final battles began on 12 June with the capture of Mount Harriet by 42 Commando, the Two Sisters by 45 Commando, and Mount Longdon by 3 Para; Tumbledown

Mountain fell to 2nd Scots Guards the following night, as did Wireless Ridge to 2 Para; the last features overlooking Port Stanley, Mount William and Sapper Hill, were taken respectively by 1/7th Gurkha Rifles and 1st Welsh Guards, the latter with two companies of 40 Commando under command, on the morning of 14 June, and later that day the

3rd Battalion paras, landed from the *SS Canberra*, during live-fire exercise on Ascension Island.

M Company, 42nd Commando, Royal Marines, posing after taking the Argentine garrison at Grytviken. Weapon at extreme left is an L7A1 MG. Commando in sweater holds an L1A1 semi-automatic rifle. In the background at the island's whaling berth is the damaged Argentine submarine *Santa Fe*.

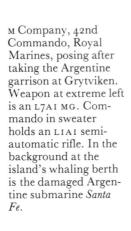

122

Argentines negotiated a general, and in all but name unconditional, surrender. Each of these battles had seen the objective secured by a combination of sustained aggression, superb junior leadership and the use of every available source of firepower, including the Milan ATGW, which was devastating in its effect on the enemy's bunkers; on occasion, too, the bayonet was used freely by the British in the close-quarter night fighting. Operation *Corporate*, as the recapture of the Falkland Islands was known, provided a textbook example in the correct use of elite forces, and its lessons will be studied for many years to come.

45 Commando passes the Two Sisters rock formation on their way to Port Stanley.

2nd Para digging in beneath Sussex Mountains.

Royal Marines advance inland on foot while equipment is transported on Snow Cats.

5th Infantry AA gunner ready for Argentine air attacks.

Royal Marine Commando in position with his L7A1 MG. The San Carlos beachhead is in the background.

Raising the Union Jack
seven weeks after the
Argentine invasion.

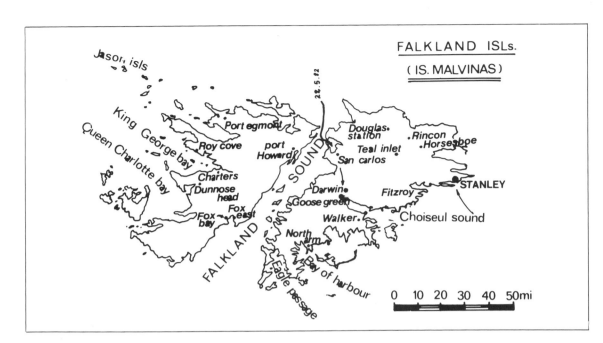

FALKLAND ISLs.
(IS. MALVINAS)

Jason isls

King George bay
Queen Charlotte bay
Port egmont
Roy cove
port Howard
Charters
Dunnose head
Fox east
Fox bay

FALKLAND SOUND

22.5.12
Douglas station
Teal inlet
San carlos
Darwin
Goose green
Walker
North arm
Bay of harbour
Eagle passage

Rincon
Horseshoe
Fitzroy
STANLEY
Choiseul sound

0 10 20 30 40 50mi

British Ships under
Argentinian attack in
'Bomb Alley'.

A British sailor (left)
and an Argentine sol-
dier of 12th Infantry
Brigade receive medi-
cal treatment on the
Canberra.

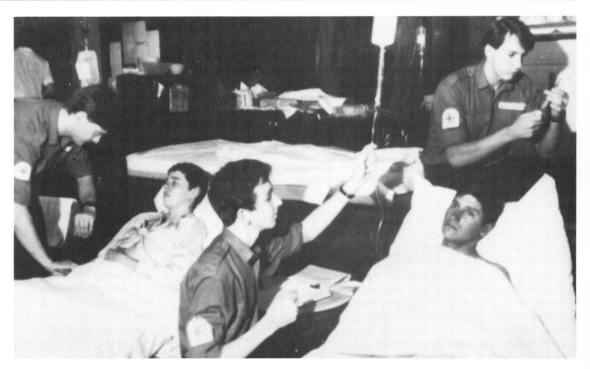

Argentinian Puccara, one of the most dangerous air threats to the British Forces.

Signing of the South Georgia surrender document on board HMS Plymouth by Captain Alfredo Astiz, watched by Captain David Pentreath (HMS *Plymouth*) and Captain Nicholas Barker (HMS *Endurance*).

Sea Harriers and Harriers GR3 provided close air support to the ground forces.

RN helicopter's Marine gunner and observer in the Falklands.

Royal Marines leaving the San Carlos beachhead.

Royal Marines entering Port Stanley, Falklands.

JUMPING INTO A HOT DZ
GRENADA 1983

Dawn, 25 October, 1983... Eighteen C-130s, escorted by AC-130E Spectre gunships, swoop low over the dark Caribbean Sea on their way to the tiny paradise island of Grenada, where life had recently turned from a daydream into a nightmare.

Leading the formation in his lizard-green Hercules was Lt. Colonel Hugh Hunter, CO 8th Special Operations Squadron. On the flight deck, peering into the darkness, was Lt. Colonel Wes Taylor, CO 1/75 Ranger Battalion. Behind him in the transports, his 300 men tensed for action. Following were 250 more elite troops led by Lt. Colonel Ralph Hagler, CO 2/75 Rangers.

The two forces had taken off some eight hours before from Ft. Stewart, Georgia (1/75) and Ft. Lewis, Washington – missioned to capture the Point Salinas airstrip on the southern tip of Grenada. Planning to land on the 'hot' runway following the paradrop of a vanguard company which would secure the area in a surprise *coup de main* attack, the Rangers were to capture the airfield quickly and move to rescue American students held hostage by Cuban forces on the island. However intelligence information which became

Troops from 1/75 and 2/75 Rangers advance inland from the airfield.

Marines outside St.
George prepare for a
patrol.

available in-flight reported heavy anti-aircraft defenses on the airstrip as well as in the
overlooking hills. It was thus decided that
everybody would jump.

Scrutinizing aerial surveillance photos
and comparing contours on the topographic
maps available, the Ranger commanders
concluded that the hostile guns would be
ineffective below 500 feet of altitude. Wishing
also to minimize the drop time and thus
reduce vulnerability to enemy fire, they
decided on a fast, ultra-low, 'Shotgun' jump
method which, in the circumstances, would
give maximum benefit to the parachutists.

A USMC AH-1T Cobra
gunship prepares for
takeoff from the *Guam*,
to support ground
forces on Grenada.

Reorganizing in mid-air for the drop, the
Rangers assisted each other in loading heavy
gear into their packs. On the assumption that
the airstrip would probably not be immediately available for follow-up cargo landings,
orders were issued to carry maximum combat loads, with priority given to weapons and
ammunition allowing the force independent
ground action for 48 hours. One trooper said
this was the heaviest rucksack he ever carried; his load was made up of his M60 machine
gun, a rifle, a .45 pistol, 1000 rounds of 7.62
ammo, and some grenades.

Some one hundred miles east of the island,
the Hercules dropped to wavetop level to
evade possible radar detection, while a few
minutes from their target, the Rangers prepared for the drop. The aircraft were lined up
in tight formation in an attempt to get everyone out as fast as possible.

The true international
spirit of Pearl Airport
is seen here, with US-
made C-130 and CH-46
parking near Soviet
AN-26 and AN-2 air-
craft.

At 0600, as the planes roared over the airfield at 500 ft, the first formation (A Company, 1st Battalion) was dropped. Using the
fast 'Shotgun' method, the Rangers jumped
from both sides of the aircraft. They encountered extremely strong reaction from the
ground which, although it did not hit the aircraft, was dangerous enough to interrupt the
drop. Mighty Spectre AC-130E gunships
were called in; the aircraft turned into a pass
over the airfield, spotted the various 23mm
gun positions on the hillside, and immediately opened fire with all their guns, engaging
them for several minutes. Spectre gunships
were hit by ground fire, but no substantial
damage was caused. After 15 minutes, the
drop was 'Go' again, with most of the air
defense and small arms fire suppressed by the

US Marines securing
the streets of St.
George while searching
for Cubans and Grena-
dian rebels.

AC-130s. The gunships continued to patrol.

At 0615, 250 Rangers from 2nd Battalion came in for the jump; Lt. Col. Hagler was the first man from 2/75 to go, hitting the ground nine seconds later. In all, it took the Rangers of the 2nd Battalion 21 seconds to hit the silk. The parachutes of two soldiers snagged; they dangled as much as 20 feet below the Hercules before jumpmasters were able to pull them back inside. The two landed later and joined the rest of the paradropped force.

The low jump was a calculated risk, decreasing total time in the air to 30 seconds and reducing the DZ to a minimum. The Rangers, used to jumping from at least 800 feet, jumped here with T 10 parachutes, older than the regular MC 1 but more immune to wind effects. They were not equipped with reserve

'chutes; after all, if trouble did arise, there would be no time for it anyway....

Despite the high-risk jump and 20 kt ground wind (over 7 kt faster than the safety margin), only one paratrooper broke his leg. Another landed with his heavy gear at sea, but swam ashore.

Quickly grouping on the small landing area, the Rangers charged the enemy with small arms, supported by the Spectre gunships flying close cover; Navy A-7E Corsair IIS and A-6E Intruders stood by in case of need.

The enemy, overwhelmed and surprised by the fast, decisive move, quickly retreated to the beach, some of them surrendering to the Rangers. Of the Cuban battalion guarding the airstrip, only one company was defending the base; the other two grouped on the beach, facing an amphibious attack they expected from the west.

In fact, all the Cuban and Grenadian gun and MG emplacements at the airfield were positioned with coastal defense in mind; their effectiveness against the incoming aircraft was therefore marginal. Located on hill slopes, the guns were in perfect position to cover the airfield perimeter facing the sea and the runway, but the American force, partly covered by the hillside, was relatively well protected.

On Grenada's northeastern side, Marines were establishing a beachhead at the coastal town of Grenville to secure the nearby Pearl Airport and the northern part of the island. (Pearl was Grenada's main international airport until the scheduled opening of the new airfield at Point Salinas in April 1984.) Before dawn, Navy SEAL (Sea-Air Landing) teams landed in fast assault boats at several points around the airport and beach-head, securing the area and maintaining static surveillance. One of the teams, however, was detected by an enemy patrol, which pinned it down throughout the night by fire from an APC.

The main force of 22 MAU (Marine Assault Unit) was lifted from the USS Guam at dawn by USMC Sea Knight CH-46 helicopters, landing at Pearl without substantial opposition in an attack coordinated with the Rangers at Salinas. USMC Cobra and USAF Spectre gunships supported the assault. While seizure of

82nd Airborne paratroopers after the battle.

These members of the Caribbean peace force, assembled for the Grenada operation, were more a political show of force than combat fighting element, but enabled the US to reduce the political risk of the operation.

133

12.7 D.Sh.K. quad-mount heavy AAMG in position on Grenada, 1983.

the airfield was relatively easy, other Marine operations were not as simple. Their attack helicopters, operating in support of both Army and Marine troops on the island, suffered several losses, with two Cobras shot down by SU-23 guns and their pilots killed; a CH-46 was also lost at sea near St. George. In fact, the determined AA fire forced the Navy and Air Force to strike more heavily than planned, and any AA site became a first priority target over any other assignment. Later, some 17 twin-barrel 23mm guns and five quad-mount DShK were captured intact.

C-130 Hercules from 317th MAW, Pope AFB, bringing supplies to the forces in Grenada.

90mm recoilless guns were among the first items airlanded by the Rangers in Salinas after the drop. They succeeded in knocking out three APCs with these weapons. Guns belonging to 2/75 Rangers are seen here with ammunition boxes and the soldiers' personal gear.

2/23mm light AA gun site located on a hilltop at the Salinas airfield which offered a clear view of the shore, weapons warehouse, and the end of the runway.

OPENING A LIFELINE

The runway itself was littered with heavy equipment and concrete debris with stakes hammered in. With the airfield perimeter almost secured but occasional sniper fire still ringing out, the Rangers began to clear the runway, opening the airfield for the incoming American cargo planes with the heavier equipment. The Rangers encountered two attacks by Cuban companies positioned beyond the airfield slopes facing the sea. This

105mm guns from the 82nd Airborne artillery battalion positioned at Pt. Salinas. Six of these guns were deployed on the island. Their main role during the first days was in supporting Ranger operations, as well as suppressing sniper fire around the airfield.

Troops from 325th Inf., 82nd Airborne Div., ready to board the C-141 jet transports at Pope airbase.

USMC CH-46 crewman operating a .30 cal. MG.

CH-53 helicopters seen after landing the hostages evacuated from Grande Anse medical school.

MAC military police performed both airport security and POW guard duties on Grenada.

US Rangers guarding Cuban detainees prior to moving them into a secured compound on the island.

force, trying to escape, ran into units from 1st Battalion and was defeated in a short fight. Another Cuban attack maneuver, supported by three BTR-60 APCs, was spotted on the eastern end of the airstrip. Unfortunately for the Cubans, they were detected by Rangers from 2nd Battalion before they were able to dismount. Every Ranger in the area fired at them with small arms, LAW, and recoilless guns. Two APCs were instantly hit, creating a death trap for the infantry inside; the third, trying to escape, was caught in a cross fire by 1st Battalion Rangers and Spectres flying above, and destroyed. Navy A-7Es came in to finish the job.

MILITARY AIRLIFT

By 1100 hours, the airfield was declared secured, and MAC C-130s and C-141 Starlifters started landing medics, evacuating wounded soldiers and bringing in supplies, operating from several air bases in the US through a forward staging base at Grantley International Airport on Barbados. The aircraft came in under sporadic gun fire and mortar bursts, sometimes circling above the airfield for hours until the fire was suppressed. Due to snipers mainly at the end of the strip but even at close quarters on the ramp, as many as five aircraft had to circle together above Salinas

at various times during the first days, capable of using only as much as 5000 feet of the 9100 foot runway. Only one C-141 and two C-130s could be parked on the ground simultaneously; the Starlifter would pull to the end of the runway and turn around while the Hercs landed behind it, pulling off onto the busy tarmac. With only one all-terrain downloader, unloading the C-141 was a lengthy process.

THE 82ND ARRIVES

First alert on the possible deployment of the 82nd Airborne was given to division staff at 0800 hours, two days before the landings. At 2030 hours the next day, the H-hour was given to the division, with orders to mobilize its Ready Brigade (DRB), using an Emergency Deployment Readiness Exercise (EDRE) as a cover. The regular standard for such an exercise in the Rapid Deployment Forces in the US (CENTCOM) is 18 hours, although 20 hours of preparation were given to the 82nd. At 1007, 25 October, the first aircraft took off, landing four hours later on the newly secured airfield at Point Salinas.

The airstrip was still under fire from sniper and mortar positions located in the surrounding woods. Although Rangers and support aircraft tried to suppress this fire, it wasn't until the third day that the area was fully secured. The build-up of the main force on the island, however, was rapid and intensive, and along with the 82nd's fast arrival, immediate demoralization was induced among the opposition; many decided to flee or gave up, with others escaping inland to attempt guerrilla operations from the jungle-covered hills.

The 82nd's main mission was to relieve the Rangers at the airfield so that they could turn to releasing the hostages taken by the Cuban and Grenadian FRA.

RESCUING THE AMERICAN HOSTAGES

American students caught by the *coup d'etat* were placed in a precarious situation – there were fears that their plight might become similar to that of the Khomeini-held diplomats in Teheran three years before. Immediate rescue action was deemed necessary.

The students were concentrated on two campuses – at True Blue Medical School near the eastern end of the Point Salinas airfield, and further north at Grande Anse near the town of St. George. Each group was guarded by about 100 armed Cubans.

At a later stage, the Marine task force sailed to the island's western shore and the 'Racetrack' landing zone near St. George. The vehicles seen here are part of the THQ assembled at the LZ.

At 1530 on 26 October, Lt. Col. Hagler's Rangers mounted a classic operation, assaulting the Cuban guards with complete surprise using Marine Sea Knight helicopters flying low towards the beach. With Spectre gunships suppressing Cuban counterfire, the Rangers, jumping from their choppers at 10 feet, stormed the campus and had the hostages free within minutes. One helicopter, hit by ground fire, crashed into the sea; the Rangers scrambled to safety through the opened ramp and took shelter from Cuban fire behind the sea wall.

Two BTR 60 APCs destroyed by the Rangers near Salinas airfield. Their crews attempted a counterattack but suffered heavy casualties.

THE FINAL MOVE

27 October – the Rangers were now ready to go home; after five days of preparations and combat, they were exhausted. The next order came like a thunderbolt: take the Edgmont military barracks, where Soviet, Cuban and Grenadian FRA were believed to have established a stronghold defended by AA guns and mortars.

CH-46 helicopters seen on the flight deck of the amphibious assault ship, *Guam.*

MEDEVAC Blackhawk from the 82nd Airborne seen at Salinas airfield. The new terminal is seen in the background.

Again, Navy A-7s and Air Force AC-130s were called in; the Rangers now decided to use four UH-60 Blackhawks from the 82nd Division in order to overcome the high barbed wire fence surrounding the target. While landing under fire in the cramped space inside the camp, one of the helicopters was hit; its pilot, wounded in the arm and leg, lost control, and his chopper tumbled into another UH-60 already on the ground. Five Rangers who jumped were killed or seriously wounded. Another Blackhawk came in to land but the pilot, shocked by the sight below, came in too fast and hit the wrecked helicopters; its whirling rotors struck more Rangers. The high price paid at Edgmont was in vain – only 12 dead Cubans were found in the camps. The rest – if there were any – escaped. This experience had a sobering effect on the Rangers and their commanders. The next day they were sent home.

Another major operation carried out that day was the seizure of Richmond Hill and its prison, where several political prisoners were being held. Also seized was the high ground overlooking St. George, the capital. The prison was taken without opposition. A Marine assault on the island of Carriacou also met with no resistance.

In early November, 22 MAU/USMC pulled out of the island for Beirut, replaced by 508/82nd at Pearl Airport and Grenville. Six thousand troops from the 82nd Airborne,

supported by some 300 soldiers from several Central American and Caribbean countries, remained in Grenada and Carriacou. A flotilla sailing from Norfolk, Virginia to exercises in the southern Caribbean was also available if needed for the defense of the island.

FIRST LESSONS

URGENT FURY gave the US armed forces another chance to prove themselves after the bitter Vietnam war and humiliating failure of the rescue operation in Iran. After several years of reorganization, planning and preparation for the unexpected, a situation arose in which the selected forces had an opportunity to prove their worth; some did well, others did not.

It is now clear that, as far as Rapid Deployment Force combat readiness is concerned, the 18-hour alert goal was fully realized. In fact, much less time was needed to lift the 82nd DRB as well as the Rangers QRF (Quick Reaction Force). Navy and Marine units also proved capable of fast reaction.

Ranger operations were excellent in both their planning and execution. Overwhelming forces brought about a rapid deterioration of the situation on the enemy side. Main-

'Agricultural equipment from Cuba' says the sign, but inside...

AC-141 B Starlifter on the runway in Grenada.
82nd Airborne troops are on guard.

82nd Airbornes, Grenada 1983.

141

taining the effect of surprise and using unconventional methods of warfare kept the initiative on the US side all the time. Exhaustive training of both Rangers and Marines proved effective in combat.

Carried away by their success, the commanders did not read the situation correctly. Pushing their forces beyond their limits took a heavy and unnecessary toll of elite units in operations easily capable of being performed by less skilled troops.

A major lack of intelligence was a key factor in the hasty planning of some operations. Although there was easy access to the various locations on the island on which lived several hundred American citizens, US intelligence was not prepared for what was described as 'a dangerous military buildup in the Caribbean.' Furthermore, the US operational intentions were learned of by Cuba, which reinforced Grenada with troops and new supplies. Real time intelligence must be available in any operation, even for low intensity battle which was expected here.

LZ arrangements and safety regulations should be considered in combat as well as in training. The chaos at Edgmont would have been much greater if one of the Blackhawks had caught fire, although on the other hand, the new helicopter's high level of survivability was proven; despite the fact that they had crash-landed and had been torn apart by one

another, the main airframe and passenger and crew compartments of the three aircraft remained intact.

Battlefield survivability was improved with the new Kevlar helmet recently issued to the US infantry units. Several helmets hit by small arms fire survived intact, while no steel helmet could have taken a similar hit.

POW factors will have to be re-evaluated for a similar contingency. A full battalion-sized force of skilled personnel had to perform POW control and could not carry out their planned role in the fighting. A less elite infantry force (without major support needs) could have done the job as well.

Facing page:
A USMC CH-46 Sea Knight transports Marines in Grenada. Note the flare dispensers in the blue box mounted aft, and the front door gunner with .30 cal. MG.

A C-141 StarLifter takes off from Salinas.

Victorious troops from 82nd Airborne pose for the camera.

82nd Airborne troops boarding UH-60A helicopters taking them on a patrol of the high inland mountains.

PRESENT DAY ELITES

Soviet landing party before embarkation.

THE SOVIET UNION

Since the end of World War II the Soviet Union has expanded its Airborne Forces, coming to possess more Airborne Divisions than the rest of the world put together. Even in peacetime these divisions are maintained at full strength and are recruited from the pick of the conscript candidates offering themselves for selection. Such importance is attached to the Airborne Forces that they operate under the control of the Commander-in-Chief of Strategic Direction and are commanded by a General of the Army, who is the equivalent in rank of the Commander-in-Chief Land Forces. In the event of an airborne operation being planned, the entire strength of the Air Force's Military Transport Aviation is immediately placed under the control of the Commander of the Airborne Forces.

Soviet airborne troops on exercises. One of the Soviet paratroopers' specific missions is the elimination of the tactical nuclear threat in the battlefield zone. Here, troops practice a surprise attack on an enemy 'missile base'.

Each of the Soviet Union's eight Airborne Divisions is approximately 7000-8000 strong, is fully mechanized and possesses tremendous firepower. The divisional order of battle consists of three parachute regiments, each of three battalions; a reconnaissance battalion; an assault gun battalion; an anti-tank battalion; a howitzer battalion; a multiple rocket-launcher battalion; an anti-aircraft battalion; a signals battalion; a transport battalion; a chemical warfare company; and an engineer company. Of the nine parachute battalions, at least five are equipped with the BMD-1 APC, and the proportion is rising. In total, the division is equipped with over 1500 vehicles.

Ostensibly, such lavishly equipped formations represent a most serious threat to NATO's strategically sensitive rear areas, but to the practical difficulties of actually dropping or air-landing a complete division in circumstances where air supremacy cannot be guaranteed must be added the high-risk nature of airborne operations generally, including the probability that dispersion will take place outside the designated dropping zone, particularly likely as a result of the Soviet preference for, by modern standards,

comparatively high level dropping. It is possible that the Russians have decided that after allowing for heavy casualties during the approach flight, in the landing itself and as a result of dispersion, insufficient troops could be rallied to seize and hold a divisional objective, and it has been suggested that they no longer anticipate divisional strength airborne operations in Western Europe. Nonetheless, there is still plenty of scope for smaller and expendable airborne battlegroups based on parachute regiments or battalions. Nor should it be forgotten that the Airborne Forces provide Russia with its own Rapid Reaction Force, capable of flying at

Soviet paratroopers boarding an AN-12 for flight to the drop zone.

The Marines disembark from APCs after landing on the beach.

Soviet paratrooper.

short notice to the far end of the Soviet empire, where relations with the People's Republic of China have occasionally degenerated into outright hostilities. Whatever role the Kremlin may have planned for its Airborne Forces, it must be admitted that during the closing weeks of the war against Japan they were handled with skill and imagination. Against this it must be said that since 1945 they have had no combat experience whatever, save in the limited war context of Afghanistan, where thus far they have failed to bring about a decision.

In a different category to the Airborne Forces are the Airborne Assault Troops, who wear the same uniform but are part of the Land Forces and not Strategic Direction troops. The Airborne Assault Troops were raised in 1969 as a result of the border clashes with China and are transported by helicopter. They are organised in brigades, each consisting of a helicopter assault regiment with 64 aircraft, a squadron of heavy-lift helicopters, and three airborne rifle battalions. Each Soviet Front at present has its own Airborne Assault Brigade which is used on the main axis of the Front's Tank Army to capture objectives behind the enemy's forward defended zone. This branch of the Soviet service is expanding and the formation of Air Assault Divisions is not improbable; this will not affect the status of the parachute-trained Airborne Forces divisions for the present, although some decline in their numbers might be expected in due course.

Although they have yet to prove themselves, the Airborne Assault Troops are regarded as an elite and are trained intensively in their role. They should not be confused with the one motor-rifle battalion in three which is trained for occasional helicopter operations.

The Soviet Army's Special Forces (SPETS-NAZ or Diversionary Troops) also wear the Airborne Forces' uniform but operate quite independently. Their function is the assassination of senior officers, the elimination of headquarters and communications centers, and general disruption and destruction within the enemy's hinterland, into which they drop by parachute or are lifted by helicopter immediately prior to a Soviet offensive. Each Army has its own 115-strong

Soviet paratroopers disembark from an AN-22. Note the Frog missiles transported by the unit.

Soviet paratroopers
boarding an IL-76
transport.

SPETSNAZ company and each Front a SPETSNAZ brigade. Recruiting is obviously highly selective and preference seems to be given to athletes and sportsmen in the international class, who possess the necessary fitness, stamina, knowledge of the countries in which they are to operate, and have a grasp of the relevant languages. Desirable though these qualities might be, they are a rough-and-ready yardstick by which to recruit Special Forces and on their own would not secure entry into, say, the British Special Air Service Regiment or the US Army's Green Berets. The quality of SPETSNAZ troops, therefore, remains an imponderable, although they doubtless gained useful experience during the suppression of the Hungarian uprising and the Soviet invasion of Czechoslovakia.

The use of sea-power as a means of spreading Soviet influence throughout the world

Parachuting and a difficult obstacle course are only part of the periodical Soviet Marine readiness exercises.

Soviet Marines overrun by PT-76 tanks.

Motorized rifle infantry soldiers operating an automatic grenade launcher.

dates from the appointment of Admiral Sergei Gorshkov, a veteran of the amphibious landings in the Crimea during World War II, as Commander-in-Chief of the Navy in 1956. Since then the Soviet fleet has expanded steadily until in size it rivaled that of the United States. This expansion was accompanied by the re-establishment of the Naval Infantry in 1961-1962, followed by the provision of tank landing ships, LCTs, and all the impediments of amphibious warfare. Each of the Soviet Navy's four fleets, Northern, Pacific, Baltic and Black Sea, possesses a 'brigade' of Naval Infantry, that of the Pacific Fleet being the largest, consisting of two tank and five motor rifle battalions and supporting artillery. The Naval Infantry has the pick of the annual conscript intake and the nature of its work ensures that it is more flexible in its approach to training than are the Land Forces, which rely on constant repetition to drive home tactical lessons. Senior Russian officers have commented that a Naval Infan-

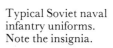
Column of Soviet BMD
APCs.

Typical Soviet naval
infantry uniforms.
Note the insignia.

Tough Soviet Marine
training includes
realistic assault on
enemy positions.

Jeeps and 57mm
assault guns are loaded
into MI-6 (HOOK)
heavy lift helicopter
during maneuvers.

57mm assault gun is
loaded on airborne
delivery pallets before
exercise.

try regimental battlegroup is capable of achieving the same results as a conventional motor rifle division.

The Soviet Navy also has its own SPETS-NAZ 'brigades', each consisting of a squadron of miniature submarines, two or three battalions of frogmen, a parachute battalion and a communications company. One such 'brigade' is attached to each fleet, although Naval SPETSNAZ units are, of course, equally capable of using vessels of the large Russian Merchant Marine and trawler fleet from which to mount their clandestine operations. They may also mount underwater guard on Soviet warships in foreign harbors, and in this context it is worth remembering that no satisfactory explanation has ever been given regarding the death of a British frogman, Lt-Commander Crabbe, in violent circumstances during a 'good will' visit by part of the

The cargo is dropped; its fall is braked by a retro-rocket charge. Heavy cargo like BMD and trucks can be dropped using this technique.

Soviet paratroops deployed on a drop zone marked with smoke and defended by a BMD-supported vanguard.

Soviet infantry disembark from MI-8 HIP assault helicopters.

Soviet fleet to the United Kingdom in the 1950s.

In structuring its elite forces the Soviet Union has made an exhaustive study of the airborne, amphibious, deep penetration and clandestine operations of the Second World War and has thoroughly digested the lessons arising from these. It is fully prepared to commit these forces during the opening stages of any major offensive against the West or against China, and to this end the Party willingly supplies the generals and admirals with a full toy-box. Yet, formidable as these forces may be, they have never fought a first-class enemy and possess only limited experience against irregular troops. In the world of elite forces, real expertise cannot be acquired from the training manuals.

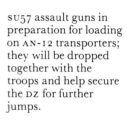

Soviet Marines briefed before attack. Behind the men are their BTR 60 amphibious vehicles.

SU57 assault guns in preparation for loading on AN-12 transporters; they will be dropped together with the troops and help secure the DZ for further jumps.

THE UNITED STATES

In sharp contrast to those of the Soviet Union, the elite forces of the West have been more or less continuously engaged in operations at varying levels of intensity for the last forty years. For example, the United States Marine Corps, America's oldest elite force, has fought major campaigns in Korea and Vietnam and has been deployed operationally in areas as widely separated as Lebanon and the Dominican Republic.

Today, the USMC is an all-volunteer force numbering 194,000 with 33,000 reserves. Training is extremely tough and demanding,

and potential recruits are warned that the right to wear the uniform will be hard-earned. At the present time the Corps consists of three active divisions, with a fourth division in reserve. The 1st Marine Division is based on the Pacific Coast; the 2nd Marine Division, assigned to the Atlantic and the Mediterranean, is based on the Atlantic Coast; and the 3rd Marine Division is based at Hawaii and Okinawa and assigned to the Pacific and Indian Oceans. Each of these divisions is capable of supporting the others in appropriate circumstances, this being

Battalion landing 3/3 Marines across the beach at Thepa, Thailand; Exercise Cobra Gold, 1983.

White Beach, Okinawa – members of the 3rd Marines Battalion guard aggressor 'prisoners' captured during Exercise Fortress Gale; 1979.

'Captured' weapons; Fortress Gale. The realism was complete, as even Soviet-made weapons (AK-47, RPG, etc.) were used.

The 2nd MAW's CH-46 helicopters during heliborne assault.

LTVP-7, the Marines' amphibious workhorse, seen here during the assault on Onslow Beach. This is the only amphibious vehicle capable of surviving a surf as high as three meters, which is essential for assault from the sea. The Soviet BTR/BMP family cannot stand up to this surf, and are used mainly for river crossing.

amply demonstrated during the Cuban Missile Crisis of 1962. They also contribute one constant-readiness unit which is permanently assigned to the United States' Rapid Deployment Force. Further constant-readiness formations known as Marine Amphibious Units (MAUs) are deployed with the Sixth Fleet in the Mediterranean, the Seventh Fleet in the Indian Ocean, and aboard ship at Diego Garcia Island, again in the Indian Ocean.

The Marines regard flexibility in deployment as being their primary asset, coupled with their ability to secure a rapid build-up. Although the USMC accounts for only 3% of all US military personnel, they nevertheless represent 15% of America's ground troops. Their high ratio of combat to support troops and their organic air support capability give them the potential to deploy quickly and effectively against areas which seem likely to pose medium-scale threats.

There is, however, another side to the coin which is less satisfactory. Marine landings today have only a fraction of the naval gunfire

Marine advances toward the target.

Aviation has played a vital role in the Marine Corps for the past 70 years.
Photo shows a F-4J Phantom.

LVTP-7 and LCM-8
amphibious landing
craft, carrying M60
tanks, approaching the
beach at Thepa, Thailand.

Korean tracked landing vehicle comes ashore during amphibious assault conducted with Marines; Exercise Team Spirit, 1982.

support that was available during World War II, an inevitable result of the replacement of the gun by the missile as the primary naval surface weapon. This problem is being given serious consideration and has led to the re-activation of the 16-inch gun battleship *New Jersey,* at present operating in support of the USMC Peace Keeping Contingent in Lebanon. Again, a lack of sufficient suitable ships and a shortfall in amphibious lift capability makes it difficult for the USMC to engage in more than one major operation simultaneously. These factors have led cost-conscious politicians to question the continued existence of the Corps in its present form. It has been suggested that the Rapid Deployment Force, relying as it does on air transport, is faster, more flexible and safer to deploy, and is therefore the logical successor to the USMC as the spearhead of American military policy overseas.

Conversely, the Reagan Administration has re-established the American global commitment and it is unthinkable that the Marines, with their high state of readiness, motivation and battlecraft, should have no part to play in such a scenario. There are numerous remote areas of the world – the Persian Gulf, the Red Sea, the Pacific and Indian Ocean atolls, the Philippines, the Caribbean, Central and South America, the southern and eastern shores of the Mediterranean, and northern Norway – where deployment of USMC units provides a more suitable response than the Rapid Deployment Force. In this context, therefore, the respective roles of the USMC and the RDF should be seen as complementary; a case, in fact, of horses for courses.

US Marine Divisions are equipped with a wide variety of weapon systems, with special emphasis placed on infantry weapons and small arms. The M16 assault rifle and the M60 machine gun, grenade launchers and other standard infantry weapons provide the backbone of the Marines' firepower, backed by such anti-tank weapons as Dragon and TOW, mounted on jeeps or other carriers. Mobility is provided by the LVTP-7, an amphibious APC protected by aluminum armor and armed with one heavy machine gun. At the present time the LVTP-7 is the only amphibi-

Landing ops during Exercise Team Spirit; 1983.

Marine squad commander awaits orders to move out.

Marines during live-fire assault exercise.

ous APC capable of negotiating the heaviest surf, outclassing both assault vehicles used by the Soviet Naval Infantry, the BTR-60 and the more modern BMP-1. With helicopter gunship and gunfire support, the LVTP-7 enables the Marines to hit the beach and drive inland before the enemy has time to coordinate his responses.

Once the division is ashore heavy fire support is provided by an M60A1 tank battalion, which is also responsible for the mobile anti-tank TOW units, and by an artillery regiment which is equipped partly with M110 self-propelled 8-inch howitzers and partly with 105mm towed howitzers. The Divisional Support Group, consisting of a Headquarters and Service battalion and an engineer battalion, is responsible for coordinating the logistic aspects of the mission, including the construction of helicopter landing pads and the provision of transport within the beach-head.

Marine from the 7th MAB digs in; Exercise Gallant Eagle, 1982.

International cooperation: Royal Marines instructor guides US Marines in small unit commando operations at Camp Lejeune; 1982.

Marine from 38th MAU stands guard during a simulated evacuation at Vieques Island, Puerto Rico; Exercise Ocean Gale, 1981. His armament consists of an M16 and M203 grenade launcher.

us Marines constantly exercise to keep up their high fighting standard. Here Marines are seen during "Ocean Venture 81" and "Display Determination 81".

101st Air-Assault soldier manning M60 MG position during winter training exercise in upstate New York.

During its assault landing and in subsequent operations the division will receive overhead protection and close air-ground support from a Marine Air Wing. This consists of a fighter attack group with 72 F-4J Phantoms, two attack groups with a total of 32 A-4M Skyhawks, 36 A-6E Intruders and 40 AV-8A Harriers, and two helicopter groups with a total of 24 UH-1N command and control ships, 16 AH-1J gunships, 72 CH-46F assault and 42 CH-53D heavy lift helicopters.

The second largest elite force possessed by the United States is, of course, its airborne arm, which is designated as the Army's strategic reserve, being capable of rapid deployment anywhere in the world. To the laurels earned by the airborne divisions in World War II have been added successful

Exercise Snow Eagle – 101st Air-Assault elements from Fort Campbell, Kentucky, provide fire cover with Dragon missiles.

Members of the 101st
Air-Assault Division
descend on ropes from
a CH-47 Chinook.

165

combat parachute drops during both the Korean and Vietnam Wars, and at the present time the famous 82nd Airborne Division, based at Fort Bragg, North Carolina, still performs this role. The division consists of nine infantry battalions, each of which includes in its order of battle 12 TOW and 30 Dragon jeep-mounted anti-tank missile launchers, heavy fire support being provided by three 18-gun artillery battalions equipped with the 105mm M102 Light Howitzer, and divisional communication, engineering and supply elements. The function of the 82nd remains the traditional one of seizing critical terrain well behind the enemy's lines and it regularly flies from the United States to Europe to participate in major exercises involving this type of operation.

The remainder of the airborne force has undergone a radical re-structuring during the last twenty years. It was the United States which pioneered the military use of the helicopter during the Korean War, and it did not take long for senior paratroop commanders, notably Major General James M. Gavin, to recognize that rotorcraft in their various forms provided a ready-made solution to some of the perennial problems of airborne warfare, and in particular answered the need for the concentrated delivery of riflemen and heavy weapons within the tactical dropping zone, accompanied by direct fire support from gunships which would suppress the immediate opposition. To distinguish the new concept from the traditional parachute drop, the troops involved were described as air assault infantry, their role being to by-pass the enemy's forward zone and then attack it from the rear. The first such formation to be raised, complete with its own attack craft, troop carriers and heavy-lift helicopters, was the 11th Air Assault Division, formed at Fort Benning in January 1963. Two years later it was re-named the 1st Cavalry Division (Airmobile) and as such acquired a distinguished combat record in Vietnam, hitting the Vietcong and the North Vietnamese Army very hard indeed wherever they might be found. So successful was the experiment that in 1968 the 101st Airborne Division was also converted to the airmobile assault role, although the stubborn

defenders of Bastogne refused point-blank to change their title.

Today, while the use of helicopters is commonplace throughout the Army, the 101st Airborne Division and the 6th Air Cavalry Combat Brigade, based respectively at Fort Campbell, Kentucky and Fort Hood, Texas, represent an elite shock element capable of either ripping a hole through the enemy's front or isolating his offensive spearheads, so disrupting his plans beyond recognition. The helicopter content apart, the 82nd and 101st Airborne Divisions are organized on broadly similar lines, although the latter has an additional artillery battalion equipped with eighteen 155mm M198 towed howitzers, which are lifted into the landing zone by CH-47 or CH-54 helicopters.

The US Army's Ranger battalions of World War II, which had originally been raised as the American equivalent of the British Commandos, were disbanded at the end of that war. During the Korean War Ranger companies were formed from Airborne volunteers, but these were incorrectly employed as

GREEN BERETS
US SPECIAL FORCES TODAY

Their world is full of contradiction: they are expert killers, but they also heal. They are the first to go into action, but the last to leave. Hard working, sometimes giving their lives for their country's interests, they get hardly any credit. Some say their work is dirty, but they say it must be done.

Only recently, with the reactivation of the 7th Special Forces Group and the Special Operations COMmand (SOCOM), have the US Special Forces, better known as the Green Berets, regained recognition for their important role in peacetime strategy.

In the aftermath of Vietnam, US defense spending was slashed to the bone. For the Special Forces, the budget was cut disproportionally, a reaction to growing public concern over the Green Berets' role in Vietnam. (Relations were not of the best between SF and regular Army men.) Of 11 groups (over 13,000 men) active in 1969, only seven remained active by 1974, and three by 1980 (about 3,000 men). The Army and Navy reduced their active units. Submarines capable of special warfare were mothballed and the Air Force forestalled procurement of special-operations-capable Talon MC-130 aircraft. The situation came to a dangerous low in 1975, with 0.1% of the total DOD budget earmarked for Special Forces and operations.

On the other side, the Soviets had expanded their Special Forces (SPETZNAZ) considerably. Operations included 'technicians' being sent to remote parts of the globe; assistance to Marxist guerrilla movements and terrorists; and the use of Cuban, East German and Arab surrogates for Third World intervention. While the Soviets recognized the peacetime importance of their Special Forces, in the US these units continued to be diminished.

As mentioned earlier, the US Special Forces reached their low point in the early and mid 1970s, with the smallest number of operational groups since the 1950s. However, since the late 70s and the Soviet involvement in Afghanistan, the US has recognized the need for rapid force projection, covert operations, and the capability either to wage low-intensity war or aid others in fighting them. The fall of Nicaragua, Angola, Ethiopia, Somalia (which later returned to the Western sphere), and Kampuchia, among others, all exemplify the determined Soviet effort to destabilize the Third World.

Today, the US is regaining some of its tools for dealing with low-intensity crisis areas. The Special Forces, if properly used, may be the answer to the slow Soviet expansion in the Third World.

A SPECIAL BREED

So you want to be a Green Beret. You can't pick up the skills through correspondence courses or survival classes in the Florida swamps. The only way is to attend the Special Forces School at Fort Bragg, N.C., and that means 16 *hard* weeks....

Tough and highly fit, all candidates are high school graduates, some with college degrees. They come from throughout the US Army with an average of three years active service behind them. The average applicant is 23, airborne-qualified and without combat experience. All he knows is what he has been taught.

The Special Forces School prepares the recruits for the tough life ahead. The school conducts a three-stage course, with individual, specialist, and team training. In the first phase, a 31-day course at nearby Camp Mackall, all the new recruits undergo a 17-hour daily schedule, seven days a week, which includes all the basic military skills. The day begins with a 45 lb backpack, six mile march, and continues with more PT (Physical Training). Only after this does the day really start, with lessons and practice in patrol, land navigation, camouflage, hostile territory survival, and living off the land. In land navigation, for example, a soldier must overcome obstacles and navigate on 12 consecutive nights in the forest. In fact, this course often comes as a shock to the soldiers. A conventional unit never teaches them survival; it provides them with food, supplies and transportation. Here, the man is on his own. For some, this proves

SF weapons specialist
armed with Soviet-
made RPK machine
gun.

US Special Forces sol-
dier practicing under-
water infiltration with
CCR1000 closed-circuit
breathing device.

to be too hard, and they may be 'terminated' by the system. The school's commander says the course is tough, and it is not shameful to fail. They try to terminate someone in what he calls a 'graceful way.' In fact, as many as 77% of those beginning the course *won't* finish (the average success rate is about 50%).

As recruits are already airborne qualified, or take a course at the nearby airborne school, they learn to use T10 and MC1 parachutes, to rappel, etc. The basic survival skills are taught by Lt. Col. Nick Rowe, a Vietnam veteran who escaped after six years of NV captivity. From his experience Rowe teaches SERE (Survival, Evasion, Resistance, and Escape) during which trainees are tested in a week-long field exercise in the Uwharrie National Forest. While the first four days of the exercise are only a 'warm-up' in the forest's southern part, the last three are held in the rough northern section, with three days of manhunts in which the soldier has to evade his pursuers. Armed only with a knife and carrying nothing but the clothes on his back, he has to run for his life, living off the land, making traps to catch small animals for food, hiding and covering his tracks, and get through these three days 'in one piece'.

During the preliminary 31 day course, the soldiers work as individuals, in contrast to other Army schools where team spirit is developed. Those who successfully pass these 31 days consider themselves lucky. But they are just beginning to learn. The course now continues with each candidate taking on a specific 'profession' in which he becomes a specialist. There are four basic skills taught in the SF:

1) *Weapons Specialists* get to know every weapon available – old, new, East, West, and Third World – from the musket loader to the 106mm recoilless rifle. They learn to operate some 85 types of weapons and to fix them with off-hand tools. Since sophisticated weapons may not always be available, they also learn to operate and construct 'home made' weapons. Heavy-weapons specialists learn about support weapons, such as mortars, RPGs, various anti-tank weapons, etc. Specialists in light- and heavy-weapons also learn each other's skills in cross-training. A weapons specialist also becomes an expert in infantry tactics on squad, platoon and company levels, including raids, ambushes, patrols, and map reading. He learns about guerrilla and counter-guerrilla

warfare, direct and indirect fire and support, security maintenance and the storage of weapons and ammunition. He also gains expertise in mine warfare, being trained with Claymore mines and booby traps.

2) *Engineer Specialists* learn to master both construction and destruction. They take an eight week course in practical sabotage. The course emphasizes 'hands-on' teaching, with much of the time spent on ranges with explosives and mines. They learn advanced techniques in the making and use of explosives, the operation of foreign-made systems, the neutralization of traps and mines, and the laying of minefields. Precision demolition, through calculations and exact measurements, is also taught on buildings, bridges and underwater targets. On the other hand, engineer candidates also learn to build houses, shelters and concrete structures. In addition, they are taught civil engineering, for use in military assistance roles in foreign countries.

3) *Communications Specialists* become experts in all kinds of communications. They learn professional, high speed Morse code and cryptography, and the

assembly and maintenance of communications equipment and stations. Upon completion of the five week academic stage, they undergo three weeks of field training, where they are parachuted 'behind enemy lines' in the Pisga National Forest. There they operate in four-man teams, maintain contact 24 hours a day with their base, move between several locations, and transmit, encrypt and decode messages.

4) *Medical Specialists* face the longest and most difficult course in the Special Forces. Candidates who have spent eight weeks in medical training prior to their preliminary course at Camp Mackall now take further courses. They spend six weeks at Ft. Sam Houston Military Medical Center in Texas, where they are introduced to the diagnosis and treatment of diseases, and the requisite medical techniques. The remainder of the seven weeks is spent in military hospitals learning from on-the-job training, and at the advanced medical training school at Ft. Bragg, where they study fragmentation and ballistic wounds, fractures and surgery, and practice mass casualty treatment. As the medical specialists are trained as physician substitutes and not

Green Berets on winter warfare training.

An SF MG 3 gunner, during Green Beret training at Ft. Bragg.

Facing page:
An SF communications specialist operates an HF portable radio, while covered by his A team member.

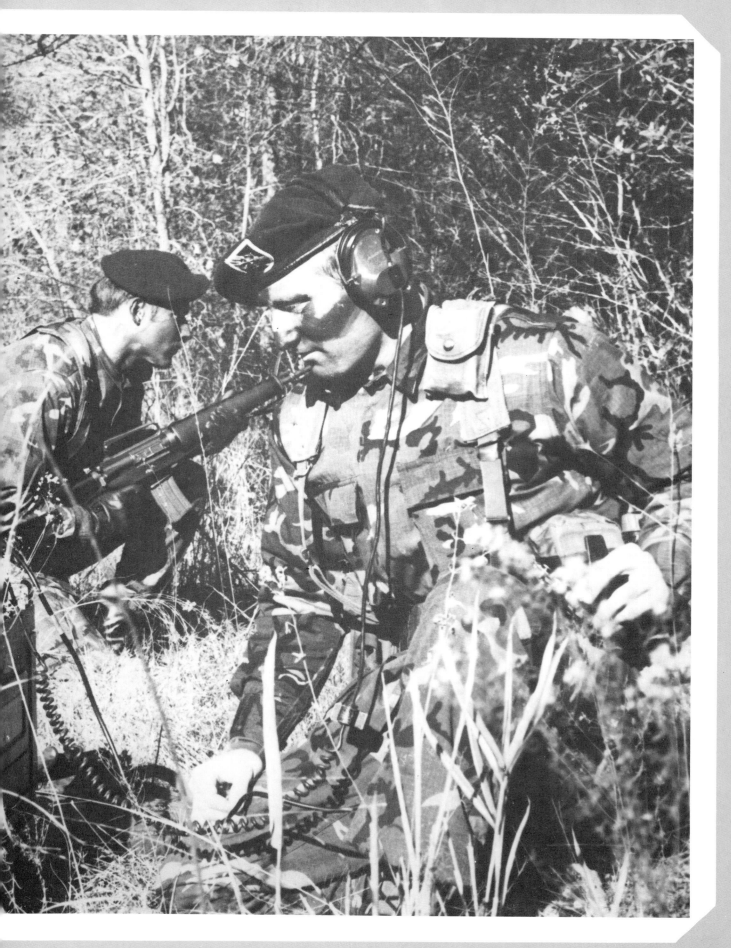

as assistants, their medical skills go far beyond those of other Army medics. Though they cannot perform surgery, their expertise can save lives in the field. Animals the candidates treat as part of their studies are regarded as patients, and if the 'patient' dies, the candidate is terminated. The medical specialists are sometimes the only medically-trained personnel that natives in a detachment area may ever see. Therefore, they are thoroughly trained in the treatment of disease, sanitation control, and civic action programs. In short, being a one-man hospital, the Special Forces medic is a 'jack of all trades'.

A fifth course in *Intelligence,* for SF-qualified NCO's, includes tactics in non-conventional warfare, interrogation, photography and fingerprinting.

In parallel to the four basic skill courses, SF officers take a two month course which prepares them for A Team command. Among the 72 topics taught are special operations techniques, guerrilla warfare, and tactics for use in enemy rear areas. They also study foreign internal affairs, the legal aspects of non-conventional warfare, cross-cultural communications, and the interrogation of prisoners. They are trained to run a Mobile Training Team, the bread and butter of the Special Forces in peacetime, in which they conduct the military training of foreign soldiers. SF officers are also introduced to psychological warfare and operations, counter-terrorist methods and crisis-management skills.

In the final five weeks of the course, the A Team itself is crystallized. The final segment begins at a preliminary 'staging base' at Camp Mackall. Teams of candidates are formed and then parachuted into the woods and swamps of Uwharrie Forest in the heart of North Carolina, far from civilization. There, they face 'aggressor' forces from the 82nd Division, a unit that feels obliged to defeat the A Teams. (This rivalry is still hot after 30 years, when the first Special Forces commander drew some 1,000 men who were the cream of the Airborne Forces, to form the newly-established SF units.) This final test is the climax of the whole course. After parachuting into the forest as they would do on a real mission, the A Team gathers about it a group of 'natives', a random selection of soldiers, none of whom have the slightest idea of what is going on. The would-be 'guerrillas' hardly know what a rifle looks like. Cooks, mechanics, and clerks, who are

motivated and enthusiastic to 'fight for their country', become a capable 'guerrilla' force in a month. The A Teams then lead them in non-conventional warfare against the 'aggressor' forces. Sometimes they are 'assisted' by sympathetic mountain men, who, after all these years, are either pro-Green Berets or pro-82nd Airborne. Contact with the wrong man could cost a trainee his head.

Graduates of the Special Forces school then go on to the various active groups based in CONUS, Panama or West Germany. They are cross-trained with at least one additional skill, and are oriented toward specific geographic regions. There are arctic specialists, jungle specialists, Arabic-speaking teams, and others oriented towards Indochina,

Central America, etc.

The Green Beret often finds himself on a completely different mission from what he expected when he volunteered for the force. Many of the teams are involved in the military assistance training of government forces, such as in Liberia, Oman, Somalia, Sudan, and Tunisia. They often deal in non-military assistance, such as civilian medical care, hygiene and sanitation, the building of houses and bridges and the paving of roads. In such non-military efforts, they win the hearts and minds of civilians and soldiers alike. This day-by-day involvement in civilian affairs is the A Team's first priority in peacetime operations, while during war, members may turn on their popular image of 'super soldiers' and 'cold-eyed

SF rappelling from a Huey helicopter at Fort Bragg.

killers'. Training for both war and peacetime missions never ceases. Every Special Forces trooper is continuously busy improving his skills or learning new ones.

To perform his non-conventional missions, the Special Forces soldier must be trained to infiltrate deep into enemy-held territory. He may parachute from high altitude at night. He may use a raft or swim under water with SCUBA equipment. He may walk, or even crawl to his destination, and might have to remain still while the enemy passes. All means of infiltration are hazardous; therefore, training is necessarily very thorough.

HELO (High Extraction Low Opening) is a parachute free-fall technique, in which a parachutist jumps from 35,000 feet equipped with oxygen and a pressure system, and opens his MC 3 parachute manually at the minimum safety level. In order to accomplish this, the parachutist stabilizes himself in the air and learns to navigate while falling. For practice in high winds, candidates train at the Wright-Patterson AFB wind tunnel to learn free fall stabilization and control. Only after that can they try a real HELO jump.

HEHO (High Extraction High Opening) is a method where the parachutist jumps from high altitude and then opens a Ram-Air gliding parachute. He navigates himself in the desired direction at a rate of 1 km/1000 ft of descent, or, in a favorable wind, at a rate of up to 3.3 km/1000 ft. Therefore, a HEHO paratrooper can land at a point between 35 to 100 km from where he jumped, crossing borders or coastlines into enemy territory, completely unnoticed and undetected. On such missions, the parachutist carries navigation equipment and more oxygen than he would if using the HELO method. Another method of infiltration is the underwater approach, using SCUBA equipment to reach a hostile coastline from a submarine or after being dropped at sea. The Special Forces are now replacing their old equipment with a new closed-breathing apparatus, the CCR 1000. This device supplies oxygen to the diver, which is purified and used again and again, and therefore does not release air bubbles which can be detected on the surface or by other divers. The CCR 1000 is usable to a depth of up to 100 fathoms. A new Gregorlar diver's suit developed for the Navy is also being used. It has a more durable design with less gauges to monitor. The complete system has adequate air supply for four hours. As it is lighter and more durable, it can be used for extended-range para-SCUBA operations.

Lt. Col. Rowe's extended SERE course is also popular. It has three levels of difficulty: the general orientation course, which is recommended to everyone 'in the trade'; a trainer's course, which is more difficult and in-depth; and a practical, highly-detailed course for high-risk personnel such as pilots, Special Forces members, etc. One of the primary roles of the course is to teach evasion and survival. Here, students learn the art of camouflage and hiding, as well as fast escape, navigation and survival.

Another short course is the sniper course, in which students learn marksmanship of the highest degree. In order to qualify, one has to achieve a multi-hit head shot from 600 meters! But marksmanship is not everything. The sniper does not have a second chance, and like a hunter, he has to be patient and remain concealed. In this course, snipers learn to identify potential targets and positions, and use shadows, foliage, and camouflage in order to survive.

THE 'A' TEAM

Whether it's swimming, mountain climbing, traversing snow, moving undetected through the jungle, parachuting from airplanes or helicopters, or paddling a raft up an alligator-infested river, the A Team is a complete unit.
The A Team is the point of the Special Forces

SF engineer prepares to detonate explosives.

sword. With only 12 men, this team of specialists can do almost anything, from blowing up a bridge to building a road or treating a sick child. The team was molded on the three-man OSS (Office of Strategic Services) team that parachuted into occupied France during WWII to organize and supply resistance fighters.

In peacetime the A Teams have the skills necessary to train guerrilla forces of up to a battalion in size, or regular soldiers of friendly governments. The team is headed by a captain and a lieutenant commanding 10 NCOs. Each is a specialist in at least one of the following skills:

1) Light/heavy weapons;
2) Communications;
3) Demolitions/engineering;
4) Medicine.

Most A Team members are cross-trained in at least two skills – on a mission, they may be separated or killed, creating a void which others could easily fill. SF officers come from regular units, retraining in the SF officer's course and taking the main course's five-week final test. After the officer's course, they take command of a team and coordinate SF elements with local (guerrilla or government) forces.

A NEW TREND

After a decade of cutbacks and 'low profile', the Special Forces are again coming into their own. With the SF's unique capabilities appreciated once more, the three services are expanding equipment procurement and authorizing more and more forces for the units. It is now expected that 1,000 more soldiers will be added to the force in a new group soon to be established.

Accordingly, all non-conventional US Army forces were consolidated with the establishment of the 1st SOCOM at Ft. Bragg. The 1st SOCOM is responsible for four SF groups: the 7th, 5th, 10th, and the new 1st Group. The two Ranger battalions are also under this command, as are the 'PSYOP' regiment-sized psychological warfare unit, and the 96th Civil Affairs (CA) Battalion. At war, their mission is to infiltrate into the enemy rear area, assist resistance forces and conduct Unconventional Warfare (UW) against the enemy. The CA is to establish contact with civilian populations within the battle area. Other SFs will be attached to regular Army units to support them in UW. In low-intensity conflicts or peacetime, the command is responsible for military assistance to over 36 countries, with 26% of the Army mobile training teams made up of Special Forces. In Central America,

Rappelling is essential for the Special Forces when operating in mountainous areas. It's also a very good way of infiltrating enemy-occupied areas.

This 5th SFG soldier rappels from an army helicopter hovering at 100 feet above the ground.

they assist guerrilla forces and the Salvadorian and Honduran armies in extensive military training. Citing Nicaragua, some critics have said that this assistance and its influence did not prevent the Sandinistas from taking power. They have also cited Liberia, where one of the trainees, Sgt. Maj. Samuel Doe, himself took power only two years later! As Doe's policies have not proved as anti-American as those of the Sandinistas, Special Forces assistance apparently can also help maintain good relations with a new regime.

A new role for the Special Forces is strategic intelligence, a by-product of their being behind enemy lines. The A Teams are now equipped with special high speed digital 'burst-transmission' systems which can evade detection by enemy scanners and can transmit information on enemy locations, order of battle, morale, etc.

Prepared to fight in any environment, SF groups are trained in Canada and Alaska for cold weather arctic conditions, and at Ft. Bragg, Korea and Sierra Nevada, Ca. for mountain warfare. They take jungle training in Panama, and

desert training worldwide. They are region-oriented; each A Team soldier knows at least one of 11 foreign languages.

The 4th Psychological Operations group is composed of three battalions which assist the three active groups. A new battalion is now being activated with the fourth SF group. Each battalion is capable of operating broadband radio stations which they deploy by air, with 500 kw transmitters and high antennae. Specialists can assemble a radio station in the jungle, far from civilization, in a matter of days. Operators and 'DJs' speaking in the local languages are available, and information can begin to be broadcast very rapidly. For further means of psychological warfare and propaganda, the group uses a high-speed printing facility for pamphlets, documents and newspapers. They can also carry a mobile printing machine with them and operate it in their deployment area. Voices can be amplified by loudspeakers installed on helicopters and jeeps, carried on the back, or operated by remote control. The Civil Affairs battalion concentrates its efforts in civilian assistance, disaster control, and other needs.

conventional line infantry and virtually fought to destruction without being able to demonstrate their full potential. Nonetheless, Korea had clearly indicated the Army's need for Ranger units, and in 1954 the 75th Infantry Regiment was raised for this purpose, the regiment inheriting the traditions and battle honor Nyitkyina from the 475th Infantry Regiment which had itself been formed from the survivors of Merrill's Marauders before being disbanded in July 1945. At present the regiment consists of two battalions, the 1st at Fort Stewart, Georgia and the 2nd at Fort Lewis, Washington. Each is specially trained to execute independent operations anywhere in the world, including raiding and long-range reconnaissance, being inserted into their operational zone

Rangers practice hand-to-hand fighting.

Basic skills such as tactics and rifle use are taught in the foreign military assistance role.

A Green Beret A Team on patrol.

Part of an A Team landing by inflatable boat.

An SF soldier can survive in the wild for a long time without support. Note his heavy back pack.

either by parachute, small boat, submarine or on foot. In any war between NATO and the Warsaw Pact it is anticipated that the Rangers will be most active between 50 and 150 kilometers behind the latter's front.

In the aftermath of World War II the Soviet Union strove to exert its influence among the uncommitted nations of the world by supporting so-called 'popular fronts' or 'liberation movements' in their attempts to replace existing regimes with a communist administration. These attempts generally took the form of a protracted guerrilla war which would destroy the morale of the indigenous security forces and erode confidence in the *de facto* government; simultaneously, the guerrillas would guarantee support for themselves by a campaign of terror among the very people they claimed to represent, ruthlessly murdering all who opposed them. It was, therefore, as a direct counter to this developing situation that the United States Army Special Forces, better known today as the Green Berets, were raised in 1952 at Fort Bragg by Colonel Aaron Bank.

The mission of the Special Forces was to beat the guerrilla at his own game and deprive him of local support. This involved a study of his methods, weapons and languages, an understanding of the local population and its problems, the ability to survive in environments which varied between the arctic, the desert and the jungle, and a high standard of general military training, including parachute jumping, since this would be the normal method of Special Forces insertion. The SF would operate in small teams which would spend long periods relying solely on their own resources and for this reason were – and remain – highly selective in their choice of recruit. The type of man wanted is already a trained regular soldier, but he must also demonstrate that he possesses stamina, initiative, imagination, self-discipline and compatibility; contrary to popular myth, there is no room at all for the psychopathic killer, since the antics of this type of individual will inevitably place his entire team in jeopardy.

Naturally, the Green Berets prefer to maintain a low profile because of the clandestine nature of their work, and they did not welcome the publicity accorded them during the Vietnam War. Yet it was in Vietnam that they fully justified their theories with their Civilian Irregular Defense Group (CIDG) program, which first taught the villagers how to defend themselves against guerrilla activity and then gradually led them over to the offensive, simultaneously expanding the scope of the program to areas not previously covered. In this way much of the countryside ceased to be a communist sanctuary and in some regions the enemy was hard pressed to maintain a presence at all.

Since Vietnam the Green Berets have maintained their expertise in guerrilla and

counter-insurgency warfare, and are at pains to point out that in the underdeveloped regions of the world their civil aid tasks do as much if not more than purely military action to combat communist subversion. They are, of course, superb guerrillas themselves and in the event of an attack by the Warsaw Pact on NATO the three Special Forces Groups based in the United States will reinforce the detachment stationed permanently in Germany, the function of which is to engage in clandestine warfare up to 300 kilometers behind the enemy's advanced units.

Green Berets' weapons specialists can operate over 80 kinds of weapons.

MAKING A RANGER

Combat-experienced soldiers know the value of tough, realistic training, and the US Rangers make sure that every Ranger recruit has the ability, skill, and endurance to accomplish all tasks.

As with many other US Army units, the Rangers were deactivated after WW II and remained inactive as special units during both Korea and Vietnam. The 75th Infantry did fight in Vietnam, but in the form of 13 separate infantry companies. Only after the war, and in light of the growing capabilities of Soviet airborne units, was the need for a fast airborne force seen. Soviet capabilities in reaching far beyond their previous limits were clearly demonstrated in the 1973 Arab-Israeli war – aerial resupply of the Arab armies went beyond the 500-mile Soviet limit previously assumed by the West.

Only in 1974 did the US Army combine these 13 separate infantry companies into the 1st (and later 2nd) Ranger Battalion, 75 Inf*. These battalions were to be the Army's elite units, a light and proficient infantry. As General Abrams, the then-Army Chief of Staff put it, a Ranger force would be 'a battalion that (could) do things with its hands and weapons better than anyone'.

The two Ranger battalions were trained to be a quick Mobile Reaction Force; to operate against aggression anywhere in the world. Their main missions were:
1) Special Operations against deep enemy targets;
2) Operations alongside conventional forces;
3) Airborne, Airmobile anti-armor operations in support of larger units;
4) Rescue operations;
5) The protection of US interests worldwide;
6) To mount a US 'Show of Force' worldwide when necessary.

On top of these main missions, the 75th Infantry was also capable of deploying as a light infantry brigade and performing as traditional infantry.

In 1975, a year after the units were commissioned, the two battalions were operational. Recruiting an experienced cadre of Vietnam veterans, 1/75 and 2/75 men conducted their own training in a 'No Bullshit' program. 'Performance oriented training' meant more emphasis on weapons and their employment, and tactics and team spirit, than on technical aspects not immediately required. This approach to soldier training became the basis for the later 'Soldier Skill Manual' which laid the foundation for army training after 1976.

When they became operational and were assigned worldwide, the Rangers became aware that, despite their excel-

With camouflage and stealth, Rangers approach a target from the back using river access.

Their goal is to surviv the cold weather and fight enemy troops in the area. They are se there to support friendly forces in com bat on a simulated defensive mission, th type which may occu in northern Norway.

178

* A third Ranger battalion was authorized in 1984.

lent training, they were unfamiliar with what would most likely be their major areas of operations – the world's deserts, mountains and jungles. An environmental acclimatization program commenced, with the Rangers training in as many different locations as possible. Near their Washington State base they used the Huckleberry Creek area on Mount Rainier for mountain training, as well as other ranges in Colorado. Winter operations were tested in Alaska, Minnesota, and Wisconsin. Desert training was undertaken at the Yakima firing range· in Washington State, and in Nevada. The Rangers trained with the Navy's SEALs in Coronado, California. Jungle exercises were carried out at the dedicated jungle training facility in Panama. Joint exercises with NATO and other allies in the Pacific added a zest to the diversified operations routines. Today, under 1st SOCOM, the Rangers are 'For Hire', anytime, anywhere.

Composed entirely of volunteers, the Ranger battalion conducts its own orientation course before accepting recruits. Having already been given basic Army Individual Training (AIT), the candidates go through a four-week Ranger Introduction Program (RIP) which raises them from AIT to the basic level required by the unit. Men unable to qualify at this level are terminated. In the early days, Ranger units were so popular that many soldiers just came and went for the experience, something that caused headaches for the commanders. The RIP filters out those who cannot meet the minimum Ranger standards. Many volunteers drop out in the first week, while about 25% are left at training's end.

One of the toughest parts of the course is the third week, which the candidates spend in the woods, constantly on the move from a simulated patrol base in an exhausting program. Here they run, crawl through the mud, are chased and shot at (over their heads, of course). Patrol leaders are regularly rotated, giving command opportunity to all students. Their performance is recorded by the instructors and analyzed by the teams back at the base. As the level of exhaustion increases and those in positions of responsibility begin to fail, candidates come to understand the dangers of exhaustion and the need for proper sleep, even on extended combat missions.

A qualified soldier in the Ranger battalion can further upgrade his skills in the US Army Ranger leadership course.

Taught at Fort Benning, the course is also open to all other qualified Army personnel. This training program is aimed at

developing the leadership skills of selected officers and NCOs under high physical and mental stress. The successful candidates, now called 'Ranger-Qualified', are fully trained to lead elite Army units. Those coming to the course from the Ranger battalions return to their units as Ranger-qualified NCOs or officers, while those coming from outside the battalions are now qualified to lead other elite infantry forces. Those successful non-Ranger candidates who are officers can now one day transfer to a Ranger battalion in a leadership capacity.

The Ranger battalions send candidates to this Ranger-qualifying course only after 'Pre-Ranger' training, a three week program that makes sure all the men it sends are prepared. Almost 100% of them successfully complete the training, many with honors.

The Ranger-qualifying course is a 58-day program, with 18 training hours a day, 7 days a week. It has a beginning phase of 19 days at the base, followed by a mountain phase of 17 days held near Dahlonega, Georgia, and a 'swamp' phase of 18 days near Eglin, Florida.

The Ranger-qualified leader has been

US Ranger.

Ranger preparing for night jump from C-141.

The Rangers assemble at the DZ with heavy equipment in their back packs to sustain them for days.

US Ranger taking up position. Note his M16's short telescopic stock used by tank troops and special units.

trained to command under stress similar to that which would be encountered in combat. He is evaluated as a small-unit leader in a series of Field Training Exercises which are conducted primarily at night, under all weather conditions. Frequent and unexpected enemy contact, reduced sleep, difficult terrain, and the constant pressure of operating within restrictive times all create the atmosphere of stress. The Ranger-qualified leader must have physical and mental endurance, and confidence to successfully complete a mission.

The Ranger qualification course is also designed to teach the student to maintain himself, his subordinates, and their equipment under arduous field conditions during the subsequent mountain and swamp phases of the course. While the first in-base phase emphasizes physical training and personal skills, an off-base exercise at Camp Darby highlights patrol techniques, parachute assaults, and small unit tactics. During the mountain phase, small unit operations are taught, and the tactically proficient students study the techniques of leading

us Rangers practice rappelling during mountain training.

Facing page:
Winter exercise in Alaska begins with a parachute drop into the cold, snow covered valley using specially modified parachutes.

squad and platoon size units, exercising control through planning and preparation, and the execution phases of all types of missions. They also learn mountaineering. The candidates are constantly challenged by severe weather, rugged terrain, hunger, fatigue, and emotional stress, all increased by day and night missions directed continuously

Rangers practice forest patrol.

Forward observer from
275th Ranger Batta-
lion.

against a simulated threat force. These missions are from two to five days in duration, and culminate in a series of combat missions where the patrols infiltrate an 'enemy's' forward security area, traverse mountains, cross the rapid Tocca river, conduct night vehicular ambushes, raid 'missile sites', and continue operations in the enemy rear for an extended period of time. Leadership is tested to the maximum; any candidate may be appointed patrol leader to lead a unit of tired, hungry Rangers on another mission.

At the termination of the mountain phase, the student moves on to the last and most demanding level, the 'swamp' phase. At Florida's Eglin AFB complex, the Ranger course conducts a 'battle' against a highly sophisticated threat force. During this time, emphasis is on the application of small-unit tactics and leadership skills in a jungle/swamp environment. This phase ends with airborne-qualified students parachuting into the area, and an airland assault by the rest. Candidates spend the remainder of the time practicing aspects of jungle warfare against the 'enemy' forces.

The results of all this training can be seen in the excellent standards shown by the Rangers in Grenada. While not all Ranger operations are revealed, their training and capabilities are promising, and they could be employed, as they were in Grenada, anywhere in the world.

US NAVY SEAL TEAMS

SEAL (Sea Air Landing) teams are US Navy tactical units commissioned to conduct Special Warfare similar to that of the Rangers, but specifically related to naval targets. They are tasked with maintaining the capability to destroy enemy shipping, harbor facilities, bridges, railway lines, and other installations in maritime areas and inland river environments.

Their operations also involve the infiltration/exfiltration of SF teams, and helping agents, guerrillas, evaders, and escapees. Among their more 'quiet' tasks are reconnaissance and surveillance of enemy installations. Peacetime jobs include low intensity, counter-insurgency operations in countries friendly to the US, missions similar to those of the Special Forces, as well as foreign military assistance in all the above-mentioned tasks.

Navy SEALs receive extensive special warfare training (similar to that of SF) in advanced demolition, communications, and weapons. As well as comprising UDTs (Underwater Demolition Teams), the SEALs must be qualified parachutists, specializing in all methods of free fall. It is this expertise, combined with basic skills ranging from Gunner's Mate to Signalman, that helps mold the SEAL into a combination Frogman, Paratrooper, and Commando. Formed on 1 January 1962 following the directives of President Kennedy, who was in favor of unconventional warfare capability (he also enthusiastically supported the Special Forces) the SEALs took part in the Vietnam war. They are now assigned to each of the US Navy fleets.

US NAVY SPECIAL WARFARE PROGRAM: (NAVSPECWAR)

The NAVSPECWAR program is involved in all aspects of special naval warfare, including training, planning, intelligence, organization, and combat. Within the US Navy wartime task to conduct sustained operations at sea, Unconventional Warfare is an essential capability, mainly for power projection in a hostile area. There are three basic components of NAVSPECWAR:
1) SEAL teams
2) UDTs (Underwater Demolition Teams)
3) SPECBOTRONs (Special Boat Squadrons).

As SEAL teams are offensive in nature, the UDTs are more defensive and are involved mainly in fleet support and amphibious operations. The boat squadrons provide the teams with transportation suited especially for their missions.

NAVSPECWAR groups are assigned to the various fleets from their two home bases at Little Creek, Va. (east coast), and Coronado, Ca. (west coast). Each group staff is assigned approximately 15 officers and 70 enlisted men, with each team assigned 25 officers and some 150 men. Currently, there are two SEAL Teams per fleet, with SEAL Team III being the most recently commissioned. Detachment units are stationed at Macrihanish (UK) and Subic Bay (Philippines). Each fleet has one SPECBOTRON, with its waterborne transportation equipment.

To qualify for the teams, all candidates, officers and enlisted men alike, undergo a rigorous entry exam. With little variation, they take the same course and endure the same physical hardships. Men 30 years of age or less may volunteer for a 25 week 'Basic Underwater Demolition/SEAL' (BUD/S) training course taught at Naval Amphibious School, Coronado. It is comprised of intensive training that develops physical and mental abilities allowing one to perform NAVSPECWAR missions under adverse conditions.

The course is conducted in three phases:
1) Nine weeks of introduction to basic UDT/SEAL equipment, swimming, obstacles, and daily PT. (A 'Hell Week' Marks the culmination of this phase with no explanation needed.)
2) Nine weeks of introduction to diving techniques, open circuit and closed-circuit SCUBA, pool and ocean swims, continued PT.

3) Eight weeks of land warfare training, small unit tactics, weapons, demolitions, hydrographic recce, communications, and basic NAVSPECWAR missions. A twenty day period on San Clemente Island concludes this phase, with live-fire and demolitions. An FTX (Field Training Exercise) in the final week tests the candidate on all the knowledge he has absorbed in the past 26 weeks.

Following graduation, NAVSPECWAR personnel receive basic parachute training at the Army's Airborne School prior to reporting to their units. Other advanced training includes courses with a Swimmer Delivery Vehicle (SDV), Special Warfare/underwater communications, and other specialities tailored to their missions. Attending many other service schools, a Navy SEAL can specialize in any subject related to any future mission.

Advanced equipment and newly developed tactics offer further advantages to NAVSPECWAR. For example, during the '60s and '70s, 20 men were required to swim through a landing area in order to map it and supply necessary information on enemy defenses, shore line obstacles, and conditions. With the new Swimmer Portable Sounding Equipment, only two men need perform this task. In a scenario, a recce team would transit from a submarine or a surface ship in an SDV, approach the target area, and conduct submerged 'recon' undetected.

Another breakthrough is stand-off mine technology. Though a highly-classified subject, the limited information available shows that this technology allows a diver to use Limpet mines with a significant target stand-off capability, thereby reducing his detection potential and improving his survivability in the face of anti-personnel countermeasures deployed around the target.

Production on a new multi-mission patrol boat (PBMM) with an improved weapons platform capability, to replace the various PBs and PBRs (Patrol Boat/River) now in use is underway. The weapons used by the SEALs and UDTs are common to those used by other US Special Warfare units, such as the M16 rifle and M60 MG, as well as specialized weapons and equipment such as rubber raiding craft, hand-held sonar gear, and SDVs. SCUBA gear is now being replaced with new closed-circuit systems using pure oxygen instead of mixed gasses. The closed circuit system leaves no bubbles, therefore reducing the detectability of a combat swimmer.

US Navy SEALs landing their boat during beach assault training.

THE UNITED KINGDOM

Royal Marines landing from an LCVP.

The armed forces of the United Kingdom include three types of units which fall within the modern definition of elite – The Royal Marine Commandos, The Parachute Regiment and the Special Air Service Regiment. Even before the Falklands War all three had long possessed an international reputation second to none, their training and operational methods being widely copied throughout the world; so much so, in fact, that the green or red beret has become the acknowledged symbol of an elite unit. It is, therefore, worth examining these methods in some detail.

In 1960 Great Britain abandoned conscription in favor of smaller but highly professional armed forces. The population of the country is relatively small in relation to its size and in times of economic depression there are always more men willing to join the armed services than are likely to be accepted.

British paras operating VIGILANT ATGMs.

This is certainly true of the Royal Marines and the Parachute Regiment, and the SAS recruit solely from the ranks of trained serv-

186

British Forces air-dropped in Port Said.

ing soldiers. It is, therefore, possible to apply a policy of selective recruitment in which unsuitable elements are progressively filtered out, and unless the potential recruit measures up physically, mentally and as an individual he has no chance of entering the service.

In this context the requirements of the Royal Marines and the Parachute Regiment are extremely high. Both run pre-recruit courses in which the potential volunteer is introduced to the service and told what is required of him. He is given a number of physical tests and a personal interview with a personnel selection officer. He is carefully observed and if he is considered unsuitable he will not proceed further; at this stage some candidates, faced with reality, will also drop out of their own volition.

The successful candidate will now join his recruit course proper. For a period he can still leave the service if he wishes to do so, and some do as the training becomes progressively tougher. At this stage the importance of the squad's senior NCO instructor cannot be over-emphasised. He is an honored figure in the British military hierarchy, a man of immense experience and insight who has the responsibility of turning the raw recruit into a trained soldier fit to take his place within a rifle company. He recognises which recruits will complete the course without difficulty and which are unlikely to complete it at all; the latter he will weed out, as quickly as pos-

sible. Between the two extremes lie the majority of the squad and it is in this area that his skilled professional judgement is most required. If, for example, a recruit finds difficulty in mastering a particular aspect of his training but shows determination in his resolve to overcome the problem, the NCO will allow considerable latitude in the matter. Again, he will recognise that any recruit can, unwittingly, pass through a difficult period of adjustment and in such circumstances the NCO privately injects a little of his own perspective to encourage the man, provided he has shown promise. When, at the end of its period of training, the squad passes out, it is inevitably smaller but it contains the men the

A British paratrooper boarding a Bell 47 helicopter used in airborne scout missions.

service wants and who have most to offer the service. Its members, now fitter, faster, healthier and harder than they have ever been, also know that they are joining a crack unit in which every officer, NCO and soldier has been through the same gruelling mill as themselves. The right to wear the green or red beret has been hard earned, as will the right to keep it, but the effort involved induces a mutual respect, a tremendous *esprit de corps* and a marked unit empathy. The two Victoria Crosses of the Falklands War were both awarded posthumously to men who deliberately sacrificed themselves for the sake of others.

Other factors which are common to the British service but which are pronounced within the Royal Marines and the Parachute Regiment are the development of personal initiative and a high standard of junior leadership. The soldier is no longer expected simply to participate in an action; he is taught to think for himself and, insofar as his own sub-unit is concerned, to make a constructive contribution in both planning and execution. Likewise, junior leaders at every level are trained to think at least one stage beyond their actual appointment.

Since the end of World War II, the Commando role has remained exclusively that of the Royal Marines, every member of which, irrespective of whether he is serving aboard Her Majesty's ships or in isolated detachments abroad, is trained to Commando standard. From start to finish his initial training takes 30 weeks, the most important phase of which falls naturally into three distinct parts: *Part 1* concentrates on the acquisition of individual military skills. The recruit is trained in the use of the rifle, machine carbine, general purpose machine gun, taught fieldcraft and camouflage and instructed in the use of map and compass.

Part 2 consolidates the recruit's infantry training and introduces him to small unit tactics in attack and defense, patrolling, signals procedure, ambush, helicopter and NBC drills, the 66mm and 84mm anti-tank weapons, and grenades.

Part 3 consists of a five-week Commando course in which the recruit learns the principles of amphibious operations, raiding, survival, rock climbing and unarmed combat.

This curriculum produces an individual who is trained well above the standard of the average infantryman, but it represents only half the effort needed to produce a Royal Marine Commando. As the course proceeds, the physical demands made on the recruit become progressively greater. The marches become longer, their pace quicker, the weight carried heavier. On level going or on a downhill slope, the soldier runs; when moving uphill he slows to a quick walk. Much of the training takes place on Dartmoor, which possesses a very similar topography to the Falkland Islands, and it was almost certainly here that the term yomp was born, its original meaning 'to eat' having been extended to 'a pace at which the ground is devoured.' Such exercises also play their part in the selection process. The man who complains if he is cold,

wet, hungry and tired is not a welcome companion; the man who, while suffering the same discomforts preserves his sense of humor, will probably do well in the service. The assault course, consisting of a series of different obstacles, is designed to test courage, agility, nerve and stamina. The obstacles are arranged in a deliberate order which prevents the body from settling into a rhythm, since different skills and internal resources are required to overcome each. All well-constructed assault courses are difficult, but that of the Royal Marines is perhaps the most formidable in the world, and recruits are expected to improve their timings for its completion as their training proceeds. In addition, the Corps also has what it calls the Tarzan Course, involving the use of ropes

Royal Marines from 40 Commando perfect their camouflage techniques during Exercise Whiskey Galore 77 held on islands off the Scottish (or is it Scotch?) shore.

and netting, in which all the obstacles are some 20-30 feet above the ground; as well as testing the recruit's nerve, the completion of this course gives him tremendous self-confidence.

Even so, the recruit is not awarded his green beret until he has succeeded in passing the following tests:

1. A nine-mile speed march carrying 36-lbs of equipment, to be completed at an average speed of one mile in ten minutes.
2. The completion of the Tarzan Course, approximately one mile in length, in a time of 13 minutes or less.
3. The completion of an endurance course, 1.5 miles long, consisting of rough, boggy country, low water-filled tunnels and deep pools, followed by a 4.5-mile run to a small-arms range, the total time allowed being 72 minutes; at the range the recruit will fire ten rounds, of which six *must* strike the target.
4. A 30-mile march by sections across Dartmoor, to be completed in 7.5 hours.

The Commando which the Marine joins on completion of his training consists of three rifle companies, a support company and a headquarters company. The rifle company contains three troops, each of three sections; the support company contains an assault engineer troop, a reconnaissance troop and a heavy weapons troop consisting of a Nilan and an 81mm mortar section; the headquarters company contains the Commando's transport, cooks and administrative personnel. After serving with his Commando for a minimum period the Marine may specialise, *inter alia*, in heavy weapons, signals or assault engineering techniques. Or, if he wishes and is considered suitable, he might be posted to the Royal Marines' Special Boat Squadron.

No 40 Commando Royal Marines on patrol during Whiskey Galore 77. Note the topographical similarities to the Falklands.

British paratroops during exercises. Note their LIAI rifles and typical red berets.

British paratroopers assembling after a jump. They are dressed in combat fatigues and wearing steel helmets. Recently, British army units were issued new lightweight Kevlar helmets.

Royal Marine raiding craft in action off the Norwegian coast during a winter exercise.

During the Falklands War the SBS was sometimes spoken of as a separate branch of service but it is, in fact, a logical element in the structure of amphibious warfare and is directly descended from the Royal Marine Boom Patrol Detachments which served with such distinction in World War II. Its tasks naturally include detailed reconnaissance of potential landing beaches and their defenses, observing and reporting on the enemy's strength and movements in the area, and also underwater attacks on enemy shipping inside defended harbors. The SBS soldier is highly qualified in a number of skills. He can, for example, 'wet-drop' to join the submarine which will take him most of the way to his objective, and he will then either swim ashore or land from a collapsible canoe containing his equipment. He is an expert in concealment and survival who avoids contact, even if the local population is friendly, as in the case of the Falklands, where SBS soldiers were put ashore fully three weeks ahead of the main British assault landing. The normal tour of duty with the Squadron is for three years, at the end of which Marines return to their Commando.

In the NATO order of battle, the 3rd Commando Brigade will reinforce the northern flank in Norway, where it trains regularly.

The Parachute Regiment has as its motto the phrase *Utrinque Paratus* – Prepared for Anything. This reflects not only its availability for deployment anywhere in the world at short notice, but also the state of mind expected of its officers, NCOs and men. The Regiment is quite clear about the type of soldier it requires. 'He must be resourceful, dis-

play initiative and be able to accept the essential self-discipline of operating as a member of a team, or as an individual. A certain robust determination is necessary. His colleagues will demand reliability from him, and his changing environment will demand adaptability; he will need to be fit, and have the stamina to stay that way.'

The Regiment's recruit course lasts 22 weeks and, like that of the Royal Marines, is progressive in its demands although it is structured a little differently. The first seven weeks are devoted to developing the recruit's individual military skills and his physical abilities and culminate in an exercise on the Brecon uplands, an environment as harsh as that of Dartmoor. Those men who have not measured up to the Regiment's standards can now either leave the Army or transfer to another regiment or corps. The remainder pass on to an intensive four-week weapon training course. During the twelfth week of the course the recruit undergoes an extremely tough Pre-Parachute Selection process, 'designed to ensure that he has the

courage, fitness, stamina and determination required of a parachute soldier.' If he fails, he can transfer to another branch of the Army. If he succeeds he will then return to Brecon for three weeks of advanced infantry training in which the emphasis is on teamwork as applied to platoon tactics. After this, he completes his training with a four-week parachute course at RAF Brize Norton in which he will make eight jumps – two from a tethered balloon and six from an aircraft, including one made at night and two in full battle equipment. He will then join his battalion and after serving for a period as a rifleman

Royal Marines winter exercise in Norway.

Royal Marines of the 3rd Commando Brigade unloading equipment from a Sea King MK.4 Helicopter during exercises in Norway.

British paratroops equipped with combat gear consisting of L7A1 MAG machine gun, L1A1 rifle, Stirling SMG, and 84 mm Carl Gustav anti-tank rocket launcher.

may go on to specialize in a technical trade such as heavy weapons or communications.

In addition to training its own recruits, the Regiment will also consider applications for transfer from trained soldiers serving in other branches of the Army, but these men must still pass through the selection process before they are accepted for final training. There is no method of back-door entry into the Parachute Regiment.

The Regiment consists of three Regular Army battalions, two of which are always at immediate readiness for employment in the parachute role while the third – which is also parachute trained – serves as an infantry battalion; there are also three Territorial Army battalions which will join Rhine Army in the event of aggression by the Warsaw Pact. Each battalion consists of four rifle companies each of three platoons; a support company containing a mortar platoon (three 81mm mortar sections), an anti-tank platoon (three ATGW sections) and an assault pioneer platoon; and a headquarters company containing the battalion's administrative personnel and transport. The 22nd Special Air Service Regiment recruits only trained soldiers, and then only after a rigorous selection procedure. The qualities which it seeks are mental as much as physical and the successful candidate will be self-reliant, self-contained, possess a sense of humor and the compatibility to work within a small team regardless of rank; most important, he will possess an iron determination to finish any task he is set, and his stamina will be well above average.

The Regiment's pre-selection process lasts three weeks, during which candidates find themselves confronted with apparently motiveless frustrations and changes of plan, all of which involve hardship and increased physical effort. These are quite intentional and the candidate's reaction to them is carefully monitored, since live operations are also subject to unexpected friction. Candidates are also required to complete a 40-mile cross-country endurance march armed and carrying a 55-lb pack; the time allowed is 20 hours.

The few men who are accepted are trained in SAS tactics, parachute jumping, combat survival, escape and evasion techniques and interrogation methods. Throughout this period the candidate knows that he is still on probation and can be returned to his parent unit at any time. If he manages to complete the course successfully he will receive the coveted SAS beret and badge. He will then specialize in developing a particular skill which can be used operationally, such as languages, demolition, climbing, boat work, CW signals, overland navigation, pistol shooting and medical care.

The SAS soldier serves initially for three years with the Regiment and can remain for further three-year tours provided he is still fit enough to do so. Service with an operational squadron can, however, push the soldier's resources to their physical and mental limits, resulting in broken health and stretched nerves. For obvious reasons, steps are taken to preserve the soldier's anonymity while he is serving with the SAS.

The Regiment is sub-divided into squadrons which, in turn, are made up of troops, but the operational sub-unit is usually a four-man team, with individual skills relevant to the mission in hand.

The sustained growth of international armed terrorism by extremist groups over the past generation shows no sign of slackening and to combat this the SAS has formed its own Counter Revolutionary Warfare Wing which makes a study of terrorist techniques and evolves ways in which to defeat them. Occasionally, the British government will respond to a request from its allies for SAS assistance, as in 1977 when a hijacked Lufthansa airliner at Mogadishu was stormed jointly by SAS personnel and members of the West German GSG-9 anti-terrorist squad, an event which effectively crippled the notorious Baader-Meinhof gang. Three years later an SAS team mounted a dramatic rescue of hostages held by terrorists within the Iranian Embassy in London, an event witnessed live by millions of television viewers. Dramatic though such events are, it is to the purely military sphere of operations that the Regiment devotes all but a fraction of its total effort. Equally, it is true to say that while the majority of nations have now formed their own highly trained anti-terrorist squads, very few of these possess the overall military capability of the SAS.

Marines training in the Hebrides.

Marines disembarking from a Wessex helicopter.

Royal Marines operating 81mm mortar.

Making paras, British-style. The hard process consists of obstacle courses, long hikes, practicing forced stretcher marches, and field training.

Westland Lynx helicopter landing anti-tank MILAN team in Germany.

OTHER WESTERN NATIONS

The French Foreign Legion has always had a reputation for tough training and strict discipline, but its operations in various parts of the world have also been marked by an adaptability and self-sufficiency which render it eligible to be called the oldest of the modern elite forces. While it wears the *kepi blanc* as a mark of honor, the Legion has striven hard to eradicate the Beau Geste image, and in fact has never sought to recruit criminals and romantics as a matter of policy – rather, like any formation faced with the realities of life, it sensibly prefers men with previous military experience or younger civilians who have an interest in and aptitude for army life. Today, recruitment is selective and the qualities sought are fitness, stamina, intelligence and initiative. The composition of those serving in the Legion's ranks has always been influenced by major international events. After World War I, for example, there was a large White Russian element, many of whom were former Czarist officers, and after World War II there was a huge German influx. At the present time the percentage of British and American citizens serving in the ranks is greater than at any time in the Legion's history, due largely to the belief that the chances of seeing active service are greater in the Legion than with their own national armies.

Yet despite the Legion's international reputation for toughness and dedication, for much of its history it was none too highly regarded in France itself, where many felt that the maintenance of a large standing mercenary force was not altogether respectable. Again, relations between the French Army in general and the Legion were never good and fists were apt to fly if the two found themselves in garrison together. These attitudes have changed somewhat since the end of the Algerian War, following which the Legion left its historic depot at Sidi Bel Abbes and re-established itself in a new home at Aubagne near Marseilles, and it is now regarded in much the same way as Americans look upon their Marine Corps, that is, as the cutting edge of overseas influence. The Legion today consists of parachute and mechanized/airmobile infantry regiments, armored cavalry units, and engineer, signal and transport elements. Its active service units are based partly in Corsica, and in those areas of Africa and the Pacific where France still has interests.

The French Army also maintains several crack divisions with specialist skills. These include the *9e Division d'Infanterie de Marine*, based at St Malo, which is trained and equipped for amphibious operations in the British Commando manner; the *11e Division Parachutiste*, much of which is stationed in Corsica; and the *27e Division Alpine*, the famous mountain troops known as *Les Diables Bleues* because of their large blue berets. Significantly, these three formations are being formed into a *Force d'Action Rapide* (FAR), which is the French equivalent of the United States' Rapid Deployment Force, and to this will be added two new formations, the *4e Division Aeromobile*, equipped with 120 anti-tank and 80 assault helicopters, and the *6e Division Legère Blindée*, a light armored division with the accent on air-portability.

Italy's elite forces are smaller than those of France and place greater emphasis on mountain warfare, there being five brigades of crack Alpini infantry, one airborne and one marine brigade.

West Germany also maintains two mountain brigades, each consisting of three infantry battalions and a light artillery battalion, and three airborne brigades, each with three battalions of parachute infantry, anti-tank and engineering troops. Nor should the German tradition of deep reconnaissance be forgotten; the *Bundeswehr's* specialists in this field are trained to a very high level and in a wide variety of skills, and invariably do well in NATO reconnaissance contests.

Other NATO member states which maintain elite forces are Turkey, with one airborne and one commando brigade; Canada, the brigade strength Special Service Force for quick reaction missions, plus one airborne infantry regiment; Greece, one para/commando brigade; Belgium, three para/commando battalions; and Portugal, one commando regiment.

German paras jumping.

French paras.

ISRAEL'S ELITE FORCES TODAY

According to international sources, there are five Israeli elite infantry brigades. Of these, two are officially identified as Golani and Givati, both of which deploy a specially trained recce unit. The former has a long and distinguished combat record while the latter, only recently reformed from a reserve outfit, was one of the best fighting units of the fledgling IDF in 1948, capturing most of the northern Negev from the Egyptian Expeditionary Force. The rest of the elite brigades are airborne formations of highly trained volunteers, and have repeatedly proved themselves in action.

After these there are the Nahal (fighting-pioneering) infantry units. Extensions of the elite Palmach strike squadrons of the pre-state Hagana underground, the Nahal formations receive first-class combat training similar to that of the airborne forces, including parachute jumps.

While the above-mentioned infantry, paratroopers, and Nahal make up the bulk of Israel's elite fighting forces, the real crack formations are the *sayarot*, or reconnaissance squadrons. These small units are composed of hand-picked volunteers and commanded by the best officers – applicants are carefully selected before undergoing a grueling training course according to the highest commando standards.

Originally there were territorial *sayarot*: *Sayeret* Shaked (Almond) operated with the Southern Command and specialized in long-range desert combat with light vehicles; *Sayeret* Egoz (Walnut) deployed with the Northern Command as a mountain unit. It fought in distinguished battles on the Golan; *Sayeret* Haruv (Carob) belonged to the Central Command and fought PLO terrorists and infiltrators in the Judean and Samarian hills and Jordan Valley.

Very little has been published on another Israeli recce unit, *Sayeret* Matkal ('Matkal' being the Hebrew acronym for GHQ). This unit comprises the pick of the IDF elite force

volunteers and should rate among the world's top-trained units. According to international sources, *Sayeret* Matkal members are specially trained for all kinds of combat, and learn, among other things, free-fall parachuting, special cross-country driving, and mountain climbing. A high rate of its men are officer material, making almost every single member a potential leader. Special weapons training is needed in order to qualify, and its fighters can operate almost any weapon of both Western and Eastern origin.

While details of their missions have never been released, it is believed that *Sayeret* Matkal or selected members have taken part in most major commando raids and anti-terrorist activities involving IDF elite forces.

TRAINING WITH GOLANI

Among the elite IDF infantry units, Golani holds a place of honor. Proving its high standards in bitter fighting, the Golani Brigade has proved over and over again that it's tops. With tough competition from the paratroopers and small but highly capable scout and special units, Golani keeps in shape with an exhaustive year-round training program. It deploys actively on the front lines, with training periods of a few months' duration in between.

One winter training period was concluded with the brigade taking part in a large combined desert exercise. Relieved from the constant pressures of peacekeeping in southern Lebanon, the Golani soldiers looked forward to this hard work as a time of relaxation. Starting the period with desert orientation and small unit exercises along with intensive PT and running to keep in shape, an integrated training program was soon begun. Exercises at company and battalion level followed. Helicopter assault, fire support, logistics, resupply, and field camp routines were carefully tested and retaught, with the aim of bringing the forces to work in perfect harmony.

Desert training is hard in the winter too. On long marches rucksacks were sometimes the only means of supply transportation. For this reason, the Brigade and IDF Logistics command are now studying simple, lightweight webbing. The increased quantities of water needed in desert environments are supplied by expandable water packs, used up before the personal canteens. Present loads are between 25 and 30 kg per soldier.

Tactically, Golani works according to the IDF combined arms doctrine. Using a typical enemy fortification line constructed by IDF engineers, Golani took part in several combined arms 'offensives' with Merkava tanks, attack helicopters, and massive artillery and air support. The soldiers defended their positions against 'counter attacks' mounted night and day in flat and mountainous terrain, using head-on and flanking moves to surprise the enemy. Golani tested Army operations books to their limits.

Among the most difficult but also most challenging parts of the training period is the 'unplanned' segment, in which the commanders are faced with the unexpected, pressed for time, and with only limited resources available. New orders and intelligence reports were issued immediately after the force completed a massive live-fire exercise. The soldiers and commanders were tired, ammunition was in short supply, there were 'casualties' to evacuate and supplies to pull in when, suddenly, everything had to be put aside and plans formulated to take on a fresh 'enemy' force recently deployed on the battlefield. This actually happens in combat, where battle plans are not always available, and 'crash programs' must sometimes be followed. This entire stage of the exercise was carried out with live ammunition and fire-support from air and artillery which had also received surprise orders.

In addition, the exercise contained a 'dry' force-on-force stage, with two units maneuvering against each other. As no winners or losers are declared, this stage is absolutely 'wild', with each of the commanders allowed to use every trick in the book (or not in the book, for that matter) to gain the upper hand.

After these extensive training periods, the brigade returns to its routines, with the highest Golani standards maintained.

Golani employs the standard IDF elite infantry brigade structure. Its battalions have each gained fame in combat (Yom Kippur on the Hermon, the Six Day War at Tel Fahar, etc.). It has a fire-support element that includes TOW units and heavy mortars, as well as engineer support units. Both operate from APC or on foot. Its famous recce unit, 'Sayeret Golani', gained fame in numerous raids behind enemy lines and while supporting the brigade at war. The Sayeret suffered a blow during the June '82 Lebanon fighting, when commander Guni Harnik and a number of his officers and men fell in battle against terrorist forces at the Beaufort Castle near Marj 'Ayoun. The fortress was eventually taken by the Sayeret. Today, the Sayeret is among the

The Golani infantrymen assault the target while covered by another team.

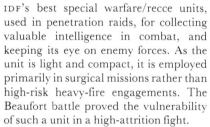

IDF's best special warfare/recce units, used in penetration raids, for collecting valuable intelligence in combat, and keeping its eye on enemy forces. As the unit is light and compact, it is employed primarily in surgical missions rather than high-risk heavy-fire engagements. The Beaufort battle proved the vulnerability of such a unit in a high-attrition fight.

Golani's 'logistical tail' is considered one of the IDF's best. In Lebanon, it was the only unit to get hot food everyday, and in combat.

Given adequate planning and dedicated men, medical and supply units were never too far from the forces, pushing their wares forward by truck, helicopter, and by parachute drops from C-130 aircraft. Ammunition, fuel, food and water, as well as spares and reserve vehicles, and even mail, were always available to the forces.

Golani troops in action.

GOLANI RECONNAISSANCE IN ACTION

Vinik squinted at his watch, which showed 0240. Looking around him in the darkness he saw the huddled figures of his men leaning exhausted on the rocks, their weapons ready. Heads nodded as the soldiers caught a few winks of sleep. In the east, over the mountain ridge, were the first glimpses of dawn; the chill penetrated to the bone, increasing the discomforts of fatigue.

The radio, squelched for silence, suddenly came alive with a quiet voice from brigade HQ reporting: 'In the vicinity of the upper ski lift there is a concentration of Syrians – go in and mop them up – out.' Vinik shook the freezing dew from his shoulders and called his men, spreading out the vanguard to lead up the narrow trail. He ordered caution, and that they should look behind each boulder for Syrians. The force set off, climbing the mountain on both sides of the track. The infantrymen, crouching with weapons at the ready, advanced quietly, spread out in loose combat formation. The sky brightened, but the morning mist still limited visibility, objects coming into view at short range. Suddenly, around the corner, a bunker with its firing aperture became visible. The silence was torn apart by machine gun fire coming from the bunker roof. Taking cover, the men returned the fire, aiming at three Syrians, clearly visible firing their weapons.

Vinik raced forward, ducking the bullets pinging on the stones as they ricocheted and whined menacingly. Approaching the bunker aperture from its flank, he threw two hand grenades into the slit, firing his Uzi submachine gun as a follow-up. He just had time to change magazines, when he was hit and fell wounded.

Someone took over, racing forward while others supported his move. Vinik was dragged to cover – but the medic could only report that he was dead. Soon after, the medic, jumping forward to attend another wounded man, was himself killed. But by then the bunker had been blown up and the three Syrians were dead as well.

Another officer came up to take command, and a doctor also came forward to attend to the wounded now collecting around the road bend, moaning quietly as they lay. Suddenly from below, engine

noise was heard as lumbering tanks ground up the narrow mountain track. As the first Centurion rounded the bend, it opened fire with a sharp crack, the 105mm round slamming into another bunker, higher up and still spitting fire.

Encouraged, Yossi and his section worked their way further up; discovering mines hidden beneath the asphalt, he started to dig them out, using his knife and shovel. As he worked feverishly, from the corner of his eye he saw a sudden ball of fire with a searing tail. 'BAZOOKA!' he yelled, but it was too late; rockets slammed into the bulldozer, igniting it, the driver scrambling from his seat. Nearby, soldiers used fire extinguishers to smother the flames, while the tank unit commander instructed his vehicles to bypass the danger. By now the 'dozer had struck a mine and was immobile, one of its tracks blown, and was blocking the road. All infantrymen now jumped off their halftracks and started storming up the hillside, mopping up the Syrians from behind the boulders; some of these, their eyes wide with fear, surrendered; others, stubbornly fighting to the end, were killed at close range. More and more Syrians gave up the fight as daylight grew.

Further up, the survivors of the recce squadron fought a desperate fight at zero range, impeding artillery support. It was up to them to overcome the hard-fighting Syrians.

Japhet tried to move forward, but stopped in his tracks as he saw the gaping hole of a well-camouflaged bunker slit. Lying flat on the ground, he looked straight into a Syrian Kalatchnikov pointed at his face. Both men were frozen by surprise, but Japhet recovered faster; grabbing the Syrian's rifle he hit him hard. The Syrian fired but missed, and the bullets tore into the sky. Then Japhet fired. The whole thing was over in seconds, but seemed to last a lifetime.

At nine, the Syrians started to give in; their fire became less accurate; more and more raised their hands. Isolated snipers were left to be rounded up as the infantry advanced to the top, exhausted but contented – they had done it again.

ARAB ELITE FORCES
JORDAN

The Jordanian special forces have been making news recently, with the Pentagon recommending them as the solution for US commitment to the Persian Gulf's security. The Pentagon recommended their expansion to a two-brigade force, fully provided with advanced equipment and improved quick-reaction capabilities.

One parachute unit existed before 1967, but remained in Jordan during the Six Day War. Today, the Jordanian special forces (commando) have already reached the size of a three-battalion force, from the original one paratroop company first established in 1963. The Jordanian special forces are handpicked; recruits are carefully selected in order to establish undoubted loyalty to the king. They are all volunteers, Bedouins who have personal tribal relations to the king. Their training is the toughest in the army, both physically and professionally. They are trained as paratroopers, for sabotage and guerrilla operations, together with the regular military professions. Later, they are sent to their various units – special (commando/paratroop) battalions. In 1969 a brigade of Special Security Groups (Saiqa)* was formed from handpicked officers and men known to be loyal to the Royal Court.

In combat, Jordanian commandos take part in elite infantry missions such as AT deployment, recce patrols, road blocks, raids, and ambushes. Their unquestioned loyalty enables the commanders to send them on peacetime missions which other soldiers may not wish to undertake. For instance, being specially trained for counter insurgency work and special operations, contingents were in action against the PLO in September 1970. Operating under command of the 4th Mechanized Division, an SSG battalion attacked Palestinian camps near the Salt-Amman highway. Another unit worked with forces blocking Syrian armor invading the kingdom from the north. Another such mission was the capture of the Inter-Continental Hotel in Amman, where terrorists were holding hostages in 1976.

The commando/para battalions (of which three are in service today) each have three companies (each with three platoons). This formation is standard in the Jordanian Army. These units are armed with various AT weapons such as Dragon ATGW, recoilless guns (106mm), mortars, and small arms (MG, M 16 rifles, etc.). They rely on jeeps and trucks for transportation, while airmobility is practiced with Alouette III helicopters, and C-130 Hercules or C-212 Aviocars for paradrops.

If these units, as an Arab RDF, have to reach the Persian Gulf, they must totally reassemble their transportation inventory, with the purchase of APCs (such as the V-300, V-150 Scout, etc.), new helicopters (UH-60 or Bell-214, and for liaison, OH-6A, Bell-206, etc.), and assault transportation which can move the total force immediately (at least 40 Hercules transports). Such a huge investment would still cost less than the effort to bring such a force from the US. But these 'special forces' would not be so special any more. In attempting to recruit more men, the Jordanians would have to deal with troops less loyal to the king. In fact, at present, the Jordanian Army is considerably oversized. With the addition of carelessly selected, highly-trained and fully-equipped rapid reaction forces in the Jordanian Army, the US might save the Gulf, but lose Jordan.

Jordanian commandos and elite armor units excelled in 1970 while fighting the PLO.

* Not to be confused with A'Saiqa, a PLO unit under Syrian Army command.

SYRIA

Among the most skilful soldiers in the Syrian army are the special commando forces, created after the Six Day War. The first paratroop battalion was raised in 1958 as an elite formation, but did not take an active part in the Six Day War battles.

These excellent elite forces are heliportable, trained to operate independently with advanced equipment. The missions of these special forces are both internal security and as elite infantry in battle. For internal security they are used like the Assad Division, to protect the regime and defend vital installations in the rear such as airports, refineries, power stations and highways.

During the opening stages of the Syrian assault on the Golan Heights in the Yom Kippur War, a mixed force from the 82nd Parachute Battalion and 1st Commando Group stormed the Israeli electronic observation post on Mount Hermon and, after a savage fight, captured it. Once reinforced, the Syrian commandos withstood repeated Israeli counterattacks, until finally withdrawn towards the end of the war when a combined attack by Golanis and paratroops recaptured the vital position.* During the post-war battles which raged over control of the mountain top in 1974, Syrian commandos clashed with Israeli paratroopers in a series of fierce battles.

Syrian commandos took part in the Lebanese civil war in 1976, when the army entered Lebanon fighting against both Christian militias and PLO units.

When the IDF invaded Lebanon in June 1982, Syrian commando units were rushed to the scene to bolster the retreating 1st Armored Division in the Beka'a Valley and central mountain sector. Israeli armored units first encountered Syrian commandos near Jezzin after a recce RPV located their position. A sharp fight ensued in which both sides showed their claws. Later, in a series of ambushes, the Syrian commando units severely mauled Israeli armored columns advancing up the Beka'a Valley, until rein-

forcements by Israeli elite infantry routed the Syrians and forced them to withdraw.

A battalion of Syrian commandos fought a courageous battle at Ein-Zakhlata in the Shouf Mountains overlooking the Beirut-Damascus main road, holding for several days against determined Israeli attacks, until most of them were killed by IDF paratroopers storming their positions.

During the cease-fire, the Syrians moved into Lebanon six more commando battalions which dug into the mountains overlooking the highway. In the battles which followed the break in the cease-fire, the commandos fought Israeli armor near Bahamdoun. Using RPG at close range, they knocked out several tanks before Golani and airborne elite units were rushed up the winding mountain roads to take them on. In a series of savage hand-to-hand clashes, the Israeli infantrymen outfought the Syrian commandos who stubbornly held on to their positions. The road was cleared only with heavy casualties, both sides demonstrating courage and determination.

The Syrian commando units are highly trained and have substantial combat experience. Commando battalions were sent to fight every possible battle in order to accumulate combat experience. This was the case in Beirut, Sidon, and Tripoli in 1976, and since then in the Shouf, the Beka'a, Beirut, Tripoli, Haleb, and Homs.

The basic formation of a Syrian commando unit is the Fug, a commando battalion made up of three elite infantry companies, with integrated medium support, AT equipment (which includes HOT, Milan, and SAGGER ATGWs), and RPG. Their personal equipment, and small and medium MGs, are suitable for night operations. Helicopter detachments can move these commando units long distances in short periods of time. Once landed, the Fug can operate independently in small elements, being less dependent on resupply than regular units.

Their toughness and training prepare them to fight in extreme conditions. Many Syrian commandos reportedly froze to death in their Lebanese mountain positions last winter, apparently too confident of their capability to survive the weather. Indeed,

* A detailed description of this action can be found on pages 90-94.

Egyptian MI-8 helicopters perform heliborne assault during Exercise Bright Star; 1982.

Syrian commandos were committed to fight to the death in the June 1982 war, which they did. If cornered, they fought it out, with few survivors. Israeli officers reported several cases of Syrian commandos killing their comrades who were trying to flee.

EGYPT

With the creation of the first paratroop unit after the Sinai Campaign, by Saad El Shazly (later COS), a serious attempt to form an elite force was made as part of the reorganization of the defeated Egyptian army. This first paratroop unit became operational in 1959. Two years later, a contingent was flown to Latakia in Syria to act as vanguard to an expeditionary force shipped to bolster Egyptian interests in the faltering United Arab Republic – the short-lived union between Syria and Egypt.

In 1963, a newly-formed battalion of paratroopers and commandos landed at Sana airport in Yemen, seizing major vantage points within the city and its environs. During the year Egyptian paratroopers and commandos were in action against the Imam's royalists, suffering heavy losses.

By 1967, the Egyptian airborne and commando force had grown to several thousand men, highly trained and motivated to act as an elite element. As an act of faith in support of the newly created Inter-Arab Command, two Egyptian commando battalions were flown to Amman on the night of 3/4 June 1967, transferred to bolster the Jordanian army. The 53rd Commando Battalion moved to the Latrun sector on the West Bank, with orders to infiltrate into the Lod airport area, while the 33rd Commando Battalion was directed to the Jenin sector further to the north, assigned to attack Ramat David airfield. Both units were routed by the Israeli attack, which came before the Egyptian commandos could act.

Egyptian commando units took part in the fighting along the Canal during the War of Attrition, attacking Israeli outposts along the Bar-Lev Line, ambushing vehicles, and placing mines along the patrol roads.

On the eve of the Yom Kippur War in 1973, the Egyptians fielded an impressive array of elite forces. These included two parachute brigades (182nd and 140th), two air assault brigades, and seven commando groups; three of them (129th, 127th, and 133rd) deployed with operational commands, while the rest were held as GHQ reserve.

Trained as special forces for the Canal crossing, the commandos acted as spearhead on 6 October, scaling the Israeli sand ramparts and setting up anti-tank ambushes with SAGGER and RPG launchers.

Under command of the 130th Marine Brigade assaulting with its amphibious vehicles over the Bitter Lakes, naval para-frogmen raided Israeli positions on the eastern shores.

Heliported commando teams were flown

into Sinai to rupture Israeli communications, assault command centers, and delay reinforcements with ambushes. Using mainly Russian Mi-8 helicopters, thirty to sixty man teams were landed behind the front line. The largest force, of battalion size, made for Ras Sudar, south of Suez, where it was spotted by IAF Phantoms and badly mauled, with most of its 250 men killed or captured before having had a chance to go into action.

Another force, landing from amphibious craft on the Romani shore, was more lucky and set up an ambush near Balousa, seriously affecting a build-up by the armor of General Adan's division – arriving at the coastal road on tank transporters. A battalion-sized Israeli force was finally ordered to mount an attack to subdue the commandos, but had to fight most of the night to overcome the enemy. Another Egyptian commando unit attacked reinforcements racing to relieve the 'Budapest' stronghold on the Tina marshes, causing heavy casualties.

In central Sinai, heliported assault teams landed during the ensuing nights and were hunted down by Israeli recce teams.

The Egyptian 182nd Paratroop Brigade, operating in the 2nd Corps sector on the canal's west bank during the closing stages of the war, fought a savage battle with Israeli paratroopers advancing towards Ismailiya under command of General Sharon. It was in these final battles that both elite forces met their match.

Egyptian MI-8 helicopter. This type is the most widely-used commando transport in the Arab and Third World armies. It is also capable of fire support. When facing the enemy in combat conditions and supported by MI-24 Hinds, these helicopters, according to the Soviet doctrine of fire suppression before landing, will attack the LZ seconds before deploying the troops.

INDEX & CREDITS

Glossary

AIT = Army Individual Training
ALC = see LCA
APC = Armored Personnel Carrier
ARVN = Army of the Republic of South Vietnam
BSU = Boat Support Units
BUD = Basic Underwater Demolition
CENTCOM = Central Command
CIA = Central Intelligence Agency
CIDG = Civilian Irregular Defense Groups
CMBS = Coastal Motor Boat Service
COPPS = Combined Operations-assault Pilotage Parties
DRB = Division Ready Brigade
ECM = Electronic Counter Measures
EDRE = Emergency Deployment Readiness Exercise
EMB = Explosive Motor Boats
FAR = Force d'Action Rapide
FLN = Front de Liberation Nationale
FLOSY = Front for the Liberation Of South Yemen
FTX = Field Training Exercise
HEHO = High Extraction High Opening
HELO = High Extraction Low Opening
IDF = Israel Defense Force
IZL = Irgun Zvai Leumi (National Military Organization)
LC = Landing Craft
LCA = Landing Craft Assault
LCT = Landing Craft Tanks
LEHI = Lochamei Herut Israel (Israel Freedom Fighters)
LRDG = Long Range Desert Groups
LVT = Landing Vehicle Tractor or Track
LZ = Landing Zone
MAB = Marine Amphibious Brigade
MAC = Military Air Command
MACV = Military Assistance Command Vietnam
MAU = Marine Amphibious Unit
MRLA = Malayan Races Liberation Army
MSF = Mobile Strike Force
MSU = Mobile Support Unit
MTT = Mobile Training Team
NAVSPECWAR = Naval Special Warfare
NLF = National Liberation Front
NVA = North Vietnam Army
PBMM = Patrol Boat, Multi-Mission
PBR = Patrol Boat, River
PLO = Palestine Liberation Organization
POW = Prisoner Of War
PRU = Provincial Recce Units
QRF = Quick Reaction Force
RAF = Royal Air Force

RIP = Ranger Introduction Program
RMBPD = Royal Marines Boom Patrol Detachment
RT = Recon Team
SAS = Special Air Service
SDV = Swimmer Delivery Vehicle
SEAL = Sea Air Landing
SERE = = Survival, Evasion, Resistance, Escape
SF = Special Forces
SLAM = Search Location Annihilation Mission
SMG = Sub Machine Gun
SOCOM = Special Operations Command
SOE = Special Operation Executive
SOG = Special Operations Group
SPECBOTRON = Special Boat Squadron
SSB = Special Services Brigade
THQ = Tactical Head Quarters
UDT = Underwater Demolition Team
UV = Unconventional Warfare
VC = Victoria Cross, or Viet Cong
VNSF = Vietnam Special Forces

Special thanks to those who gave a hand in the preparation of this book:

Charles Messenger
Lt. Cdr. Joleen Keefer
US DOD Audiovisual Dept, especially Mrs. Spriggs
Capt. Havrilla, 9th Infantry Div.; U.S.A.
Maj. Roy, 101st Air Assault Div.; U.S.A.
Maj. Curter, 1ste SOCOM; U.S.A.
Capt. Roni, IDF Spokesman's office
Herzel Kunsari, IDF spokesman's office
Maj. Baird; US Marines 2nd Div.

And to the many others who made the publication of this book possible.

Photo credits and acknowledgments:

All Hands Magazine (US Navy) p. 184, 185
Bell Helicopters p. 44, 50, 53
Brenda Ralph Lewis p. 24, 25, 26, 27
Bundesministerium der Verteidigung (German MOD) p. 195
Camera Press p. 9, 20, 29, 34, 35, 91, 92, 102, 118, 152, 153,155
Champion P. Eshel Dramit (C) p. 112
Defence Update Magazine p. 84, 146, 147, 150, 151, 154, 155
Eshel T. p. 3, 10, 11, 54, 94, 97, 159, 161, 162, 163, 188, 189, 190, 191
Gutman N. p. 57, 64, 66, 80, 82, 83, 88, 90, 91
Haganah Archive p. 58, 59
IAF p. 84, 85, 86
IDF Archives p. 62, 63, 68, 69, 70, 71, 72, 73, 74, 75, 76, 77, 78, 79, 82, 83, 88, 92, 93
IGPO p. 59, 60, 61, 67, 70, 72, 87, 96, 97, 98, 113, 114
IDF Spokesman p. 98, 99, 114, 187
Imperial War Museum p. 8, 9, 11, 12, 13, 14, 15, 16, 17, 18, 19, 20, 21, 22, 23, 28, 29, 30, 38
London Express Agency p. 110, 111
Novosti Press Agency p. 31, 33, 144, 147, 150
Royal Marines MOD p. 10, 130, 191, 193
Salamander books p. 154
Soldier Magazine p. 104, 105, 186, 187
South African MOD p. 115, 116
Tass p. 148
UKLF/MOD p. 2, 187, 191
UK MOD p. 117, 119, 121, 122, 123, 124, 125, 126, 127, 188, 189, 190, 191, 193
US ARMY: p. 45, 46, 47, 48, 49, 51, 164, 167
9th Inf. Div. p. 175, 178, 179, 180, 182, 183
101st Div. p. 165, 166
US DOD p. 6, 40, 41, 108, 109, 131, 132, 133, 134, 135, 136, 137, 138, 139, 140, 142, 143, 156, 157, 160, 161, 162, 163, 168, 175, 181
US Marines p. 36, 37
US Special Forces p. 169, 170, 171, 172, 173, 174, 175, 176, 177